Praise for *Science Friction*

"You may disagree with Michael Shermer, but you'd better have a good reason—and you'll have your work cut out finding it. He describes skepticism as a virtue, but I think that understates his own unique contribution to contemporary intellectual discourse. Worldly-wise sounds wearily cynical, so I'd call Shermer universe-wise. I'd call him shrewd, but it doesn't do justice to the breadth and depth of his inspired scientific vision. I'd call him a spirited controversialist, but that doesn't do justice to his urbane good humor. Just read this book. Once you start, you won't stop."
  —Richard Dawkins, author of *The Selfish Gene* and *A Devil's Chaplain*

"It is both an art and a discipline to rise above our inevitable human biases and look in the eye truths about how the world works that conflict with the way we would like it to be. In *Science Friction,* Michael Shermer shines his beacon on a delicious range of subjects, often showing that the truth is more interesting and awe-inspiring than the common consensus. Bravo."
  —John McWhorter, author of *The Power of Babel* and *Losing the Race*

"Michael Shermer challenges us all to candidly confront what we believe and why. In each of the varied essays in *Science Friction,* he warns how the fundamentally human pursuit of meaning can lead us astray into a fog of empty illusions and vacuous idols. He implores us to stare honestly at our beliefs and he shows how, through adherence to bare reason, the profound pursuit of meaning can instead lead us to truth—and how, in turn, truth can lead us to meaning."
  —Janna Levin, author of *How the Universe Got Its Spots*

"Whether the subject is ultra-marathon cycling or evolutionary science, Michael Shermer—who has excelled at the former and become one of our leading defenders of the latter—never writes with anything less than full-throttled engagement. Incisive, penetrating, and mercifully witty, Shermer throws himself with brio into some of the most serious and disturbing topics of our times. Like the best passionate thinkers, Shermer has the power to enrage his opponents. But even those who don't agree with him will be sharpened by the encounter with this feisty book."
  —Margaret Wertheim, author of *Pythagoras' Trousers*

Also by Michael Shermer

The Science of Good and Evil

In Darwin's Shadow:
The Life and Science of Alfred Russel Wallace

The Skeptic Encyclopedia of Pseudoscience
(general editor)

The Borderlands of Science

Denying History

How We Believe

Why People Believe Weird Things

Science Friction

SCIENCE *FRICTION*

Where the Known
Meets the Unknown

MICHAEL SHERMER

AN OWL BOOK

Henry Holt and Company • New York

Owl Books
Henry Holt and Company, LLC
Publishers since 1866
175 Fifth Avenue
New York, New York 10010
www.henryholt.com

An Owl Book® and 🛗® are registered trademarks of
Henry Holt and Company, LLC.

For further information on the Skeptics Society and *Skeptic* magazine,
contact P.O. Box 338, Altadena, CA 91001, 626-794-3119;
fax: 626-794-1301; e-mail: skepticmag@aol.com. www.skeptic.com

Library of Congress Cataloging-in-Publication Data
Shermer, Michael.
    Science friction: where the known meets the unknown /
Michael Shermer.—1st ed.
        p. cm.
    Includes index.
    ISBN-13: 978-0-8050-7914-2
    ISBN-10: 0-8050-7914-9
    1. Science—Philosophy. 2. Science—Miscellanea. I. Title.
Q175.S53437 2005
501—dc22
                                        2004051708

Henry Holt books are available for special promotions and
premiums. For details contact: Director, Special Markets.

Originally published in hardcover in 2005 by Times Books

First Owl Books Edition 2006

Designed by Victoria Hartman

Printed in the United States of America

10 9 8 7 6 5 4 3 2

To Pat Linse

For her steadfast loyalty,

penetrating intelligence,

and illustrative originality

Contents

IV / Science and the Cult of Visionaries

Introduction

Why Not Knowing

Science and the Search for Meaning

THAT OLD PERSIAN TENTMAKER (and occasional poet) Omar Khayyám well captured the human dilemma of the search for meaning in an apparently meaningless cosmos:

> *Into this Universe, and* Why *not knowing,*
> *Nor* Whence, *like Water willy-nilly flowing;*
> *And out of it, as Wind along the Waste,*
> *I know not* Whither, *willy-nilly blowing.*

It is in the vacuum of such willy-nilly whencing and whithering that we humans are so prone to grasp for transcendent interconnectedness. As pattern-seeking primates we scan the random points of light in the night sky of our lives and connect the dots to form constellations of meaning. Sometimes the patterns are real, sometimes not. Who can tell? Take a look at figure I.1. How many squares are there?

The answer most people give upon first glance is 16 (4 x 4). Upon further reflection, most observers note that the entire figure is a square, upping the answer to 17. But wait! There's more. Note the 2 x 2 squares.

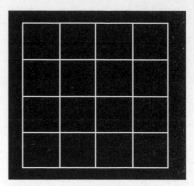

Figure I.1

There are 9 of those, increasing our count total to 26. Look again. Do you see the 3 x 3 squares? There are 4 of those, producing a final total of 30. So the answer to a simple question for even such a mundane pattern as this ranged from 16 to 30. Compared to the complexities of the real world, this is about as straightforward as it gets, and still the correct answer is not immediately forthcoming.

∽

Ever since the rise of modern science beginning in the sixteenth century, scientists and philosophers have been aware that the facts never speak for themselves. Objective data are filtered through subjective minds that distort the world in myriad ways. One of the founders of early modern science, the seventeenth-century English philosopher Sir Francis Bacon, sought to overthrow the traditions of his own profession by turning away from the scholastic tradition of logic and reason as the sole road to truth, as well as rejecting the Renaissance (literally "rebirth") quest to restore the perfection of ancient Greek knowledge. In his great work entitled *Novum Organum* ("new tool," patterned after, yet intended to surpass, Aristotle's *Organon*), Bacon portrayed science as humanity's savior that would inaugurate a "Great Instauration," or a restoration of all natural knowledge through a proper blend of empiricism and reason, observation and logic, data and theory.

Bacon was no naive utopian, however. He understood that there are significant psychological barriers that interfere with our understanding of the natural world, of which he identified four types, which he called idols: *idols of the cave* (peculiarities of thought unique to the individual that distort how facts are processed in a single mind), *idols of the marketplace* (the limits of language and how confusion arises when we talk to one another to express our thoughts about the facts of the world), *idols of the theater* (pre-existing beliefs, like theater plays, that may be partially or entirely fictional, and influence how we process and remember facts), and *idols of the tribe* (the inherited foibles of human thought endemic to all of us—the tribe—that place limits on knowledge). "Idols are the profoundest fallacies of the mind of man," Bacon explained. "Nor do they deceive in particulars . . . but from a corrupt and crookedly-set predisposition of the mind; which doth, as it were, wrest and infect all the anticipations of the understanding."

Consider the analogy of a swimming pool with a cleaning brush on a long pole, half in and half out of the water—the pole appears impossibly bent; but we recognize the illusion and do not confuse the straight pole for a bent one. Bacon brilliantly employs something like this analogy in his conclusion about the effects of the idols on how we know what we know about the world: "For the mind of man is far from the nature of a clear and equal glass, wherein the beams of things should reflect according to their true incidence; nay, it is rather like an enchanted glass, full of superstition and imposture, if it be not delivered and reduced." In the end, thought Bacon, science offers the best hope to deliver the mind from such superstition and imposture. I concur, although the obstacles are greater than even Bacon realized.

For example, do you see a young woman or an old woman in figure I.2?

This is an intentionally ambiguous figure where both are equal in perceptual strength. Indeed, roughly half see the young woman upon first observation, and half see the old woman. For most, the young and old woman image switches back and forth. In experiments in which subjects are first shown a stronger image of the old woman, when shown this ambiguous figure almost all see the old woman first. Subjects who are initially exposed to a stronger image of the young woman, when shown this ambiguous figure almost all see the young woman first. The metaphoric extrapolation to both science and life is clear: we see what we are programmed to see—Bacon's idol of the theater.

Idols of the tribe are the most insidious because we all succumb to them, thus making them harder to spot, especially in ourselves. For example, count the number of black dots in figure I.3.

The answer, as you will frustratingly realize within a few seconds, is that it depends on what constitutes a "dot." In the figure itself there are no black dots. There are only white dots on a highly contrasting background

Figure I.2

Figure I.3

that creates an eye-brain illusion of blinking black-and-white dots. Thus, in the brain, one could make the case that there are 35 black dots that exist as long as you don't look at any one of them directly. In any case, this illusion is a product of how our eyes and brains are wired. It is in our nature, part of the tribe, a product of rods and cones and neurons only. And it doesn't matter if you have an explanation for the illusion or not; it is too powerful to override.

Figures I.4 and I.5.   The 3-D impossible crate

Figure I.4, the "impossible crate," is another impossible figure. Can you see why?

All of our experiences have programmed our brains to know that a straight beam of wood in the back of the crate cannot also cross another beam in the front of the crate. Although we know that this is impossible in the real world, and that it is simply an illusion created by a mischievous psychologist, we are disturbed by it nonetheless because it jars our perceptual intuitions about how the world is supposed to work. We also know that this is a two-dimensional figure on a piece of paper, so our sensibilities about the three-dimensional world are preserved. How, then, do you explain figure I.5, a three-dimensional impossible crate?

This is a real crate with a real man standing inside of it. I know because the man is a friend of mine—the brilliant magician and illusionist Jerry Andrus—and I've seen the 3-D impossible crate myself. Like other professional magicians and illusionists, Jerry makes his living creating interesting and unusual ways to fool us. He depends on the idols of the tribe operating the same way every time. And they do. Magicians do not normally reveal their secrets, but Jerry has posted this one on the Internet and shown it to countless audiences, so as a lesson in willy-nilly knowing, figure I.6 provides the solution to the 3-D impossible crate.

Buried deep in our tribal instincts are idols of recognizable importance to our personal and cultural lives. As an example of the former, note the striking feature in the photograph from Mars in figure I.7, taken in July 1976 by the *Viking Orbiter 1* from a distance of 1,162 miles, as it was photographing the surface in search of a viable landing site for the *Viking Lander* 2.

The face is unmistakable. Two eye sockets, a nose, and a mouth gash form the rudiments of a human face. What's *that* doing on Mars? For decades this feature (about a mile across), as well as others gleaned from eager searchers perceptually poised to confirm their beliefs in extraterrestrial intelligence, claimed it was an example of Martian monumental architecture, the remnants of a once-great civilization now lost to the ravages of time. Numerous articles, books, documentaries, and Web pages breathlessly speculated about this lost Martian civilization, demanding that NASA reveal the truth. This NASA did when it released the photograph of the "face" taken by the Mars Global Surveyor in 2001, seen in figure I.8.

Figure I.6.   The 3-D impossible-crate illusion explained. Camera angle is everything!

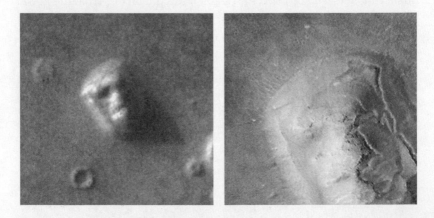

Figure I.7 (*above left*).   An unmistakable feature on Mars photographed by a NASA spacecraft in 1976.

Figure I.8 (*above right*).   The "face" on Mars morphs into an eroded mountain range.

In the light of a high-resolution camera, the "face" suddenly morphs into an oddly eroded mountain range, the product of natural, not artificial, forces. Erosion, not Martians, carved the mountain. This silenced all but the most hard-core Ufologists.

Such random patterns are often seen by humans as faces, such as the "happy face" on Mars "discovered" in 1999 (figure I.9). If astronomers were romantic poets would they find hearts on distant planets, like the one in figure I.9, also from Mars?

We see faces because we were programmed by evolution to see the expressions of those most powerful in our social group, starting with imprinting on the most important faces in our sphere: those of our parents.

We also see at work Bacon's idols of the cave in the peculiarities of religious icons that often make their appearances in the most unusual of places, such as the famous "nun bun" discovered by a Nashville, Tennessee, coffee shop owner in 1996. The idea of seeing a nun's face in a pastry provokes laughter among most lay audiences (it was featured on David Letterman's show, for example). But some deeply religious people flocked to show their reverence when the nun bun was put on display at the Bongo Java coffee shop. (An attorney representing Mother Teresa forced the bun's owner, Bob Bernstein, to remove her name from the icon.)

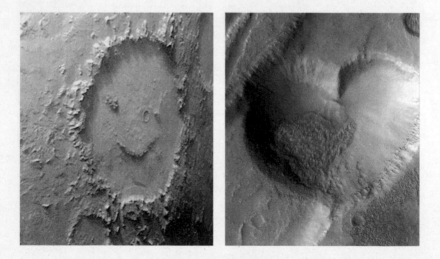

Figure I.9. The "happy face" on Mars, along with a Martian heart

Arguably the greatest religious icon in history (after the cross, of course) is the Virgin Mary, who has made routine appearances around the world and throughout history. In 1993, for instance, she appeared on the side of an oak tree in Watsonville, California, a small town whose population is 62 percent Mexican-Americans and whose dominant religion is Catholicism. In 1996 the Virgin Mary manifested on the side of a bank building in Clearwater, Florida. Once again, believers gathered around the icon, often in wheelchairs and on crutches, in some cases hoping to be healed.

A Christian group purchased the building in order to preserve the religious image, fencing off the parking lot, which is now chockablock full of candles lit in veneration. However, as I discovered in visiting the site in 2003, it turns out that there are several Virgin Marys on the sides of this building, appearing wherever there happens to be a sprinkler and a palm tree. The water, contrary to the name of the city, is not so clear. In fact, it is rather brackish, loaded with minerals that stain windows such as these (see figure I.10; the palm tree that originally stood in front of the window where the big Virgin Mary image appears has since been cut down by the owners).

The image is a striking example of the power of beliefs to determine perceptions. Instead of saying, "I wouldn't have believed it if I hadn't seen it," we probably should be saying, "I wouldn't have seen it if I hadn't believed it." As with faces, we see religious icons because we were programmed by history and culture to see those features most representative of those institutions of great power, starting with the religion of our parents.

⸎

Nowhere are such idols harder to see in ourselves than the subtle psychological biases we harbor. Consider the *confirmation bias*, in which we look for and find confirmatory evidence for what we already believe and ignore disconfirmatory evidence. For my monthly column in *Scientific American* I wrote an essay (June 2003) on the so-called Bible Code, in which the claim is made that the first five books of the Bible—the Pentateuch—in its original Hebrew contain hidden patterns that spell out events in world history, even future history. A journalist named Michael Drosnin wrote two books on the subject, both *New York Times* bestsellers, in which he claimed in the second volume to have predicted 9/11. My analysis was very skeptical of this claim (I told him in a personal letter that it would

Figure I.10 (*above left*).   The Virgin Mary on the side of a bank building in Clearwater, Florida. (The author, in the middle, is bracketed by Richard Dawkins, left, and James Randi.)

Figure I.11 (*above right*).   Another Virgin Mary on another side of the bank building

have been nice if he had alerted everyone to 9/11 *before* the event instead of after). He wrote a letter to the magazine (and had an attorney threaten them and me with a libel suit), which they published. In response, I received a most insightful letter from John Byrne, a well-known comic book writer and illustrator of Spider-Man and other superheroes. I reprint it here because he makes the point about this cognitive bias so well.

> Reading Michael Drosnin's response to Michael Shermer's column on the Bible "code" and its ability to accurately predict the future, I could not help but laugh. I have been a writer and illustrator of comic books for the past 30 years, and in that time I have "predicted" the future so many times in my work my colleagues have actually taken to referring to it as "the Byrne Curse."
>
> It began in the late 1970s. While working on a Spider-Man series titled "Marvel Team-Up" I did a story about a blackout in New York. There was a blackout the month the issue went on sale

(six months after I drew it). While working on "Uncanny X-Men" I hit Japan with a major earthquake, and again the real thing happened the month the issue hit the stands.

Now, those things are fairly easy to "predict," but consider these: When working on the relaunch of Superman, for DC Comics, I had the Man of Steel fly to the rescue when disaster beset the NASA space shuttle. The *Challenger* tragedy happened almost immediately thereafter, with time, fortunately, for the issue in question to be redrawn, substituting a "space plane" for the shuttle.

Most recent, and chilling, came when I was writing and drawing "Wonder Woman," and did a story in which the title character was killed, as a prelude to her becoming a goddess. The cover for that issue was done as a newspaper front page, with the headline "Princess Diana Dies." (Diana is Wonder Woman's real name.) That issue went on sale on a Thursday. The following Saturday . . . I don't have to tell you, do I?

My ability as a prognosticator, like Drosnin's, would seem assured—provided, of course, we reference only the above, and skip over the hundreds of other comic books I have produced which featured all manner of catastrophes, large and small, which did not come to pass.

In short, we remember the hits and forget the misses, another variation on the confirmation bias.

In recent decades experimental psychologists have discovered a number of cognitive biases that interfere with our understanding of ourselves and our world. The *self-serving bias*, for example, dictates that we tend to see ourselves in a more positive light than others see us: national surveys show that most businesspeople believe they are more moral than other businesspeople. In one College Entrance Examination Board survey of 829,000 high school seniors, 0 percent rated themselves below average in "ability to get along with others," while 60 percent put themselves in the top 10 percent. This is also called the "Lake Wobegon effect," after the mythical town where everyone is above average. Lake Wobegon exists in the spiritual realm as well. According to a 1997 *U.S. News and World Report* study on who Americans believe are most likely to go to heaven, for example, 60 percent chose Princess Diana, 65 percent thought Michael Jordan, 79 percent selected Mother Teresa, and, at 87 percent, the person most likely to go to heaven was the survey taker!

Experimental evidence of such cognitive idols has been provided by Princeton University psychology professor Emily Pronin and her colleagues, who tested a generalized idol called "bias blind spot," in which subjects recognized the existence and influence in others of eight different specific cognitive biases, but they failed to see those same biases in themselves. In one study on Stanford University students, when asked to compare themselves to their peers on such personal qualities as friendliness, they predictably rated themselves higher. Even when the subjects were warned about the "better-than-average" bias and asked to reevaluate their original assessments, 63 percent claimed that their initial evaluations were objective, and 13 percent even claimed to be too modest! In a second study, Pronin randomly assigned subjects high or low scores on a "social intelligence" test. Unsurprisingly, those given the high marks rated the test fairer and more useful than those receiving low marks. When asked if it was possible that they had been influenced by the score on the test, subjects responded that the *other* participants were negatively influenced, but not them! In a third study in which Pronin queried subjects about what method they used to assess their own and others' biases, she found that people tend to use general theories of behavior when evaluating others, but use introspection when appraising themselves; but in what is called the "introspection illusion," people do not believe that others can be trusted to do the same. Okay for me but not for thee.

The University of California, Berkeley, psychologist Frank J. Sulloway and I made a similar discovery of an "attribution bias" in a study we conducted on why people say they believe in God, and why they think *other people* believe in God. In general, most people attribute their own belief in God to such intellectual reasons as the good design and complexity of the world, whereas they attribute others' belief in God to such emotional reasons as it is comforting, gives meaning, and they were raised to believe.

∽

Such biases in our beliefs do not prove, of course, that there is no God (or Martians or Virgin Marys); however, in identifying all these different factors influencing (and often determining) what it is we see and think about the world, it calls into question how we know *anything*. There are many answers to this solipsistic challenge—consistency, coherence, and correspondence being just three devised by philosophers and epistemologists—

but for my money there is no more effective Reliable Knowledge Kit than science. The methods of science, in fact, were specifically designed to weed out idols and biases. Some patterns are real and some are not. Science is the only way to know for sure.

Cancer clusters are a prime example. As portrayed in Steven Soderbergh's 2000 film *Erin Brockovich*, staring Julia Roberts as the buxom legal assistant cum corporate watchdog, lawyers can strike a financial bonanza with juries who do not understand that correlation does not necessarily mean causation. Toss a handful of pennies up into the air and let them fall where they may, and you will see small "clusters" of pennies, not a perfectly random distribution. Millions of Americans get cancer—they are not randomly distributed throughout the country; they are clustered. Every once in a while, they may be grouped in a town where a big industrial plant owned by a wealthy corporation has been dumping potentially toxic waste products. Is the cancer cluster due to the potentially toxic waste, or is it due to random chance? Ask a lawyer and his clients hoping for a large cash settlement from the corporation, and they will give you an unambiguous answer: cluster = cause = cash. Ask a scientist with no stake in the outcome and you will get a rather different answer: cluster = or ≠ cause. It all depends. Additional studies must be conducted. Are there other towns and cities with similar correlations between the chemical waste product and the same type of cancer cluster? Are there epidemiological studies connecting these chemicals to that cancer? Is there a plausible chemical or biological mechanism linking that chemical to that cancer? The answers to such questions, usually in the negative, often come long after juries have granted plaintiffs large awards, or after corporations grow weary and financially drained fighting such suits and opt to settle out of court.

A similar problem was seen in the silicon breast implant scare of the late 1980s and early 1990s. I distinctly recall the advertisements placed in the *Los Angeles Times* by legal firms, alerting any women with silicon breast implants that they might be entitled to a significant financial award if they exhibit any of the symptoms listed in the ad, which was a laundry list of aches and pains connected to a variety of autoimmune and connective tissue diseases (as well as the vagaries of everyday life). A hotline was also established: 1/800-RUPTURE. Women responded . . . in droves, and the litigant attorneys paraded their clients in front of the courthouse with placards that read WE ARE THE EVIDENCE. In 1991, one of these women, Mariann Hopkins, was awarded $7.3 million after a jury determined that her ruptured silicone breast implant caused a connective tissue disease. Within weeks, 137 lawsuits were filed against the manufacturer, Dow Corning. The next year another woman, Pamela Jean Johnson, won $25 million after a jury linked to her implants connective tissue disease, autoimmune responses, chronic fatigue, muscle pain, joint pain, headaches, and dizziness, even though the scientists who testified for the defense said her symptoms amounted to nothing more than "a bad flu." By the end of 1994, an unbelievable 19,092 individual lawsuits had been filed against Dow Corning, shortly after which the company filed for bankruptcy.

In the end, the confirmation bias won out and Dow Corning had to pay $4.25 billion to settle tens of thousands of claims. The only problem was, there is no connection between silicone breast implants and any of the diseases linked to them in these trials. After multiple independent studies by reputable scientific institutions in no way connected to either the corporation or any of the litigants, the *Journal of the American Medical Association*, the *New England Journal of Medicine*, the *Journal of the National Cancer Institute*, the National Academy of Science, and other medical organizations declared that this was a case of "junk science" in the courtroom. Dr. Marcia Angell, the executive editor of the *New England Journal of Medicine*, explained that this was nothing more than a chance overlap between two populations: 1 percent of American women have silicone breast implants, 1 percent of American women have autoimmune or degenerative tissue diseases. With millions of women in each of these categories, by chance tens of thousands will have both implants and disease, even though there is no causal connection. That's all there is to it.

Why, then, in this age of modern science, was this not clear to judges and juries? Because Bacon's idols of the marketplace dictate that scientists and lawyers speak two different languages that represent dramatically different ways of thinking. The law is adversarial. Lawyers are pitted against one another. There will be a winner and a loser. Evidence is to be marshaled and winnowed to best support your side in order to defeat your opponent. As an attorney for the prosecution it doesn't matter if silicone *actually* causes disease, it only matters if you can convince a jury that it does. Science, by contrast, attempts to answer questions about the way the world really works. Although scientists may be competitive with one another, the system is self-correcting and self-policing, with a long-term collective and cooperative goal of determining the truth. Scientists want to know if silicone really causes disease. Either it does or it does not.

Marcia Angell wrote a book on this subject, *Science on Trial*, in which she explained how "a lawyer questioning an epidemiologist in a deposition asked him why he was undertaking a study of breast implants when one had already been done. To the lawyer, a second study clearly implied that there was something wrong with the first. The epidemiologist was initially confused by the line of questioning. When he explained that no single study was conclusive, that all studies yielded tentative answers, that he was looking for consistency among a number of differently designed studies, it was the lawyer's turn to be confused." As executive editor of the *New England Journal of Medicine*, Angell recalled that she was occasionally asked why the journal does not publish studies "on the other side," a concept, she explained, "that has no meaning in medical research." There is "another side" to an issue only if the data warrant it, not by fiat.

The case of Marcia Angell is an enlightening one. She opens her book with a confession: "I consider myself a feminist, by which I mean that I believe that women should have political, economic, and social rights equal to those of men. As such, I am alert to discriminatory practices against women, which some feminists believe lie at the heart of the breast implant controversy. I am also a liberal Democrat. I believe that an unbridled free market leads to abuses and injustices and that government and the law need to play an active role in preventing them. Because of this view, I am quick to see the iniquities of large corporations." What's a disclaimer like this doing in a science book? She explains: "I disclose my political philosophy here, because it did not serve me well in examining the breast implant controversy. The facts were simply not as I expected they would be. But my most fundamental belief is that one should follow the evidence wherever it leads." Francis Bacon would approve.

∽

My favorite example of the beauty and simplicity of science is the case of Emily Rosa and her experiment on Therapeutic Touch. In the 1980s and early 1990s, Therapeutic Touch (TT) became a popular fad among nursing programs throughout the United States and Canada. The claim is that the human body has an energy field that extends beyond the skin, and that this field can be detected and even manipulated by skilled TT practitioners. Bad energy, energy blockages, and other energetic problems were said to be the cause of many illnesses. TT practitioners "massage" the human energy field, not through actual physical massage of muscles, tendons, and tissues (which has been shown to have therapeutic value in reducing muscular tension and thereby stress), but by waving their open palms just above the skin. This is touchy feely, without the touchy. A professional nurse friend of mine name Linda Rosa became alarmed at the outrageous claims being made on behalf of TT, as well as the waste of limited time and resources in nursing programs and medical schools on it. TT was even being employed in hospitals and operating theaters as a legitimate form of treatment. In September 1994, the U.S. military granted $355,225 to the University of Alabama at Birmingham Burn Center for experiments to determine if TT could heal burned skin.

One day Linda was watching a videotape of TT therapists practicing their trade when her ten-year-old daughter, Emily, got an idea for her fourth-grade science project. "I was talking to my mom about a color separation test with M&Ms," Emily explained. As she watched the TT tape, she reports, "I was nearly 'mesmerized' by seeing nurses like my mom waving their hands through the air. I asked her if I could test TT instead (and just eat the M&Ms)." Of course, Emily had neither the time nor the resources to carry out a sophisticated epidemiological study on a large population of patients to determine if they got well or not using TT. Emily just wanted to know if the TT practitioners could actually detect the so-called human energy field. It is obvious that in waving a hand above someone's body one might *imagine* sensing an energy field. But could it be detected under blind conditions?

Emily's experiment was brilliantly simple. She set up a card table with an opaque cardboard shield dividing it into two halves, with Emily sitting on one side and the TT practitioner on the other side. Emily cut two small holes on the bottom of the board, and through each the TT subject put both arms, palms up. Emily then flipped a coin to determine which of the subject's hands she would hold her hand above. Figure I.12 shows a diagraph of Emily's experimental protocol.

With this simple design we can see that, just by guessing, anyone being tested should be able to detect Emily's hand 50 percent of the time.

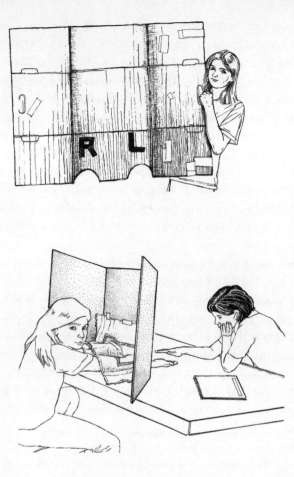

Figure I.12.   Emily Rosa puts Therapeutic Touch to the test.

Assuming that the TT practitioners can really detect a human energy field, they should do better than chance. (In fact, most said they could detect it 100 percent of the time, and in preexperimental trials without the cardboard, they had no problem sensing Emily's energy field.) Emily was able to test twenty-one Therapeutic Touch therapists, whose experience in practicing TT ranged from one to twenty-one years. Out of a total of 280 trials, the TT test subjects got 123 correct hits and 157 misses, a hit rate of 44 percent, below chance!

We published preliminary results of Emily's experiment in a 1996 issue of *Skeptic* magazine, but our readers were already skeptical of Therapeutic Touch so this generated no great controversy. But in the April 1, 1998, issue of the prestigious *Journal of the American Medical Association* Emily's complete results were published in a peer-reviewed scientific article, and suddenly Emily found herself on the *Today Show*, *Good Morning America*, all the nightly news shows, NPR, UPI, CNN, Reuters, the *New York Times*, the *Los Angeles Times*, and many more (over a hundred distinct media stories about Emily and her experiment appeared within weeks). We brought Emily to Caltech to present her with our Skeptic of the Year award, where she was also recognized by the Guinness Book of World Records as the youngest person ever to be published in a major scientific journal.

Did Emily prove that Therapeutic Touch is a sham therapy? Well, it depends on how her results are placed in a larger context. Emily did not directly test whether people are healed or not using TT, so we can only indirectly make this inference. But it is obvious that if TT practitioners cannot even detect the so-called human energy field, how can they possibly be "massaging" it for healing purposes? In any case, the "Emily Event," as it has now become known among skeptics, serves as a case study in how science can be conducted simply and cheaply (Emily spent under $10 for her entire project), and can get to the answer of an important question.

೮∽

We have seen that science is a great Baloney Detection Kit. But can it identify nonbaloney? Can it find real patterns? Of course. Smoking causes lung cancer—no question about it. HIV causes AIDS—almost no one doubts it. Earth goes around the sun. Earthquakes are caused by plate tectonics and continental drift. Plants get their energy from the sun through photosynthesis, and we get our energy by eating plants and animals that eat plants. This much is true, and much more. Yet there are true patterns that are counterintuitive. Quantum physics is one of these. It's weird, really weird. Electrons go around the nucleus of an atom like a planet goes around the sun—we've all seen the schematic diagram; only it isn't true. Electrons are quite real, like itty-bitty planets, but they are everywhere and nowhere in their orbits at the same time—identify their position and you cannot know their momentum; track their momentum and you have lost their position (this is the so-called uncertainty principle). And electrons don't exactly go "around" the nucleus; they form a "shell" described by a wave equation. And sometimes the electrons jump from inner orbits to outer orbits—and vice versa—without passing in between. How can that be? It seems impossible, but it is true nonetheless. A century of experimental evidence has been accumulated, often by physicists who were hard-core skeptics, demonstrating beyond almost any doubt that quantum physics is real.

Evolution is another of those scientific truths with which a number of people still struggle. Leaving out the religious objections (which are irrelevant in a scientific discussion), the counterintuitive nature of evolution is caused by two problems: (1) our propensity to favor the experimental sciences (e.g., physics) over the historical sciences (e.g., archaeology) as reliable sources of truth claims, and (2) our intuitive grasp of time frames on the order of months, years, and decades of a human life, in contrast to the history of life that spans tens and hundreds of thousands of years, and even millions, tens of millions, and hundreds of millions of years. These time frames are so vast that they are literally inconceivable. So let's consider a very conceivable problem: explaining the great diversity of "man's best friend," dogs. There are hundreds of breeds of dogs, ranging in size from Chihuahuas to Great Danes, and varying in behavior from poodles to pit bulls. Where did all this canine variation come from? In 2003 geneticists found out—comparing the DNA of dogs and wolves, they determined that every breed of dog from everywhere on Earth came from a single population of wolves living in China roughly fifteen thousand years ago. Imagine that—in that short span of time (*very* short on an evolutionary timescale, speeded up, of course, by human intervention through artificial selection) we have this almost unimaginable degree of variation in dogs. It doesn't seem possible, but there it is.

The human evolution story is much the same. In the May 11, 2001, issue of the journal *Science*, in a report on the "African Origin of Modern Humans in East Asia," a team of Chinese and American geneticists sampled 12,127 men from 163 Asian and Oceanic populations, tracking three genetic markers on the Y chromosome. What they discovered was that every one of their subjects carried a mutation at one of these three sites that can be traced back to a single African population some 35,000 to 89,000 years ago. The finding corroborates earlier mitochondrial DNA (mtDNA) studies, along with fossil evidence, that every one of us can trace our ancestry to Africa not so very long ago. Looking around at the incredible diversity of humans on the globe, this seems nearly impossible. But the evidence supports it, and when it does scientists change their minds. One of the chief proponents of the theory that human groups (so-called races) evolved independently in separate regions (after a much earlier exodus out of Africa), the University of California, Berkeley, anthropologist Vince Sarich, after examining the new data, confessed: "I have undergone a conversion—a sort of epiphany. There are no old Y chromosome lineages [in living humans]. There are no old mtDNA lineages. Period. It was a total replacement." In other words, in a statement that takes great intellectual courage to make, Sarich admitted that he was wrong.

Although scientists should probably make such admissions more often than they do, as a profession we are more open to error admission than most others. Can you imagine the shocked response a leading Democratic politician would hear if he made a statement like the following? "After examining the evidence for the claim that more gun control laws will reduce gun violence, I have changed my position. I am now against gun control." Or think of the stunned silence one might hear in a church if a preacher uttered this sentence: "I have carefully considered all the arguments for and against God's existence. If I am to be intellectually honest I must confess that the evidence for God is no better than the evidence against, so I hereby declare myself an agnostic." Every once in a great while such conversions happen, but they are so rare as to be newsworthy.

As Bacon so adequately argued four centuries ago, and as cognitive psychologists have experimentally demonstrated the past couple of decades, the facts never speak just for themselves. Science is a very human enterprise. Nevertheless, like democracy, it is the best system we have so we should use it to its fullest extent and apply it wherever we can in our quest to know why.

෴

This précis on knowing and not knowing introduces this volume, a collection of fourteen research articles and personal essays that I have written over the past decade—most (but not all) published in various journals and magazines (but none appearing in my other books)—about how science operates under pressure, during controversies, under siege, and on the precipice of the known as we peer out in search of a ray of light to illuminate the unknown. I have grouped them into four general sections, each of which embodies science on the edge between the known and the unknown, in that fuzzy shadowland that offers a unique perspective on both knowing and not knowing, and how science is the best tool we have to discern which is which.

Part I, "Science and the Virtues of Not Knowing," begins with chapter 1, "Psychic for a Day," a first-person account of an amusing and enlightening experience I had spending a day as an astrologer, tarot card reader, palm reader, and psychic medium talking to the dead. I did this on invitation from Bill Nye (the "science guy") for an episode of his television series, *Eyes of Nye*. With almost no experience in any of these psychic modalities, I prepared myself the night before and on the plane flying to the studio, then improvised live-to-tape in studio, managing to completely convince my sitters that I had genuine psychic powers, reducing several subjects to tears when we "connected" to lost loved ones. It was at this point that I realized the emotional impact that psychics can have on believers, and the immorality of the entire process and industry that has built up around these claims.

Chapter 2, "The Big 'Bright' Brouhaha," presents in narrative form the results of an empirical study I conducted (I include data charts as well) on the skeptical movement and the attempt to unite all nonbelievers, agnostics, atheists, humanists, and free thinkers under one blanket label—The Brights—and why, in my opinion, all such attempts will ultimately fail. The movement began at the Atheist Alliance International convention in 2003, and I was the first to sign the petition. The "Bright" movement gained momentum when myself and such luminaries as the evolutionary theorist Richard Dawkins and the philosopher Daniel Dennett came out of the skeptical closet through opinion editorials. The reaction was swift and merciless—almost no one, including and especially nonbelievers, agnostics, atheists, humanists, and free thinkers, liked the name, insisting that its elitist implications, along with the natural antonym "dim," would doom us as a movement. The entire episode afforded a real-time analysis of how social movements evolve.

Chapter 3, "Heresies of Science," presents six heretical ideas that promise to shake up everything we have come to believe about the world:

The Universe Is Not All There Is, Time Travel Is Possible, Evolution Is Not Progressive, Oil Is Not a Fossil Fuel, Cancer Is an Infectious Disease, and The Brain and Spinal Cord Can Regenerate. With each heresy, I consider the belief it is challenging, the alternative it offers, and the likelihood that it is correct. Since this is cutting-edge science, my conclusions are necessarily provisional, as in most of these claims the data are still coming in and resolution is not final. In the case of the first two claims, it may be some time before we can detect other universes, and unless we receive visitors from the future, time machines will likely remain the staple of science fiction for some time to come (pardon the pun). Evidence is mounting, however, in support of the fact that evolution is purposeless (and designless as well, at least from the top down—evolution is a bottom-up designer), that some forms of cancer are caused by infectious viral agents, and that parts of the brain and spinal cord can regenerate under certain limited conditions. As for oil not being a fossil fuel, here we would be wise to be skeptical of the oil skeptics. Even though the proponent of this claim is a renowned scientist, reputation in science only gets you a hearing. You also need reliable data and sound theory.

Chapter 4, "The Virtues of Skepticism," is a brief history of skepticism and doubt, the relationship between science and skepticism, and the role of skeptics in society. This essay began as a tribute to my friend and colleague—the venerable Martin Gardner, one of the fountainheads of the modern skeptical movement—but, as I try to do in nearly all of my writings, I also impart larger lessons for what we can learn about how science works through examining how it doesn't work. In this case, I examine skepticism itself, with some embarrassment for my own lack of initiative and insight for not doing this in 1992 when we founded the Skeptics Society and *Skeptic* magazine. In selecting these names for the society and magazine, one would imagine that we would have carefully thought out their linguistic and historical meaning and usages, but it was not until the late Stephen Jay Gould wrote the foreword to my book *Why People Believe Weird Things*, in which he discussed the meaning of skepticism, that I got to thinking about what precisely it is we are doing when we are being skeptical.

Part II, "Science and the Meaning of Body, Mind, and Spirit," begins with chapter 5, "Spin-Doctoring Science," by demonstrating how science gets spin-doctored during explosive controversies, such as the anthropology wars over the true nature of human nature. This is an in-depth analysis of the fight among scientists over the proper interpretation of the Yanomamö people of Amazonia. Are they the "fierce" people, as one anthropologist called them, in constant battles with one another over precious resources, or are they the "erotic" people, as another anthropologist labeled them, passionately loving and sexual? The answer is yes, the Yanomamö are the erotic fierce people, or the fiercely erotic people. They are, in fact, people, just like us in the sense of possessing a full range of human emotions, and a complete suite of human traits, together comprising our nature as *Homo sapiens*.

In chapter 6, "Psyched Up, Psyched Out," I utilize my experiences in the 1980s as a professional bicycle racer, particularly my founding of and participation in the three-thousand-mile nonstop transcontinental Race Across America—the ultimate test in the sport of ultramarathon cycling—to consider the power of the mind in sports, what we know and do not know about its role in athletic performance, and what this tells us about the interaction between mind and body. Since I was an active participant in the race, hell-bent on winning as much as the next guy, I entered the fray not as an objective scientist curious about whether this or that diet or training technique or new technology worked, but as a competitor in search of an edge. The farrago of nonsense I encountered along the way ultimately led to my becoming a skeptic because athletes are especially superstitious and vulnerable to outlandish claims, and I was among the gullible at this stage in my life.

Chapter 7, "Shadowlands," is the most personal commentary in the book, as I recount the story of a ten-year battle with cancer I helped my mother to wage against brain tumors (to which she eventually succumbed), and what I learned about the limitations of medical science, the hubris of medical diagnosis and prognosis, the lure of alternative medicine, and the interface of science and spirituality. I was with my mom every step of the way, from initial diagnosis of depression in a psychiatrist's office, to CAT scans and MRIs, to the surgery waiting room (numerous times), to her final hospital stay, nursing home, and, finally, when she breathed her last. How even a hardened skeptic deals with death, particularly that of someone as close as a parent, demonstrates, I hope, that skepticism is more than a scientific way of analyzing the world; it is also a humane way of life.

Part III, "Science and the (Re)Writing of History," begins with chapter 8, "Darwin on the *Bounty*," by demonstrating how science can be put to good use to solve a historical mystery—what was the true cause of the mutiny on the *Bounty*—and how evolutionary theory provides an even deeper causal analysis of human behavior under strain. Historians operate at a proximate level of analysis, searching for immediate causes that triggered a particular historical event. Evolutionary theorists operate at an ultimate level of analysis, searching for deeper causes that underlie proximate causes. In this case I am not disputing what historians have determined happened to the crew of the *Bounty*, and what in the weeks and months preceding the rebellion led them to take such drastic action against their commander. What I am looking for is an explanation in the hearts of the men, so to speak; not just how, but why, in the sense of what in human nature could lead to such actions. As such, in this study I am extending what I have done in my previous work as a professional historian. In *Denying History*, I analyzed the claims of the Holocaust deniers and demonstrated with rigorous science how we know the Holocaust happened. In my biography of Alfred Russel Wallace, *In Darwin's Shadow*, I employed several theories and techniques from the social sciences to ground my subject in a deeper level of understanding of human behavior. I am not attempting to revolutionize the practice of history, so much as I am trying to add to it the tried-and-true methods of science, so often neglected (because of the balkanization of academic departments) by historians.

Chapter 9, "Exorcising Laplace's Demon," applies the modern sciences of chaos and complexity to human history, showing how meaningful patterns can be teased out of the apparent chaos of the past. I have been thinking about how to apply chaos and complexity theory to human history since those sciences came on the scene in the late 1980s, while I was earning my doctorate in history. I have always been interested in the philosophy of history. Our flagship journal is *History and Theory*, and that title says a lot. History is the data of the past. But data without theory are like bricks without a blueprint to transmogrify them into a building. It is theory that binds historical facts into a cohesive and meaningful pattern that allows us to draw deeper conclusions about why (in the deeper sense) things happen as they do.

In a related analysis, chapter 10, "What If?," I employ the always enjoyable game of "what if" history, suggesting that scientists, too, can play this game to useful ends to explore what might have been and what had to be. Here I am most emphatically doing history not just for history's sake, although that is part of it, but for us. I believe that history is primarily for the present, secondarily for the future, and tertiarily for the past. It is great fun to ponder how the history of the United States might have unfolded after 1863 had General McClellan not received ahead of time General Lee's plans for the invasion of the North. Lee most likely would have been victorious at Antietam/Sharpsburg, which would probably have led the British and French to recognize the South as a sovereign nation, which would have encouraged them to aid the South in breaking the North's blockade of ships bringing valuable resources from Europe, which possibly would have led to the South's ability to carry out war for many more years to come, which most probably would have led to Northern war weariness and perhaps a ·congressional decision to let their Southern brothers and sisters secede from the union. Such speculation, however, is far more than merely amusing; it serves as an object lesson in historical causality: if this, then that; if not this, then not that. And this and that object lesson may serve us well for both present and future.

Chapter 11, "The New New Creationism," picks up the anti-evolution movement in America with the mid-1990s development of "Intelligent Design Theory," or ID, in which these quasi-scientific thinkers shed the cloak of the old creationists from the 1960s and 1970s who demanded a literal biblical interpretation of scientific findings, and that of the new creationists of the 1980s who were more flexible in adapting the findings of science but still insisted on a divine hand in nature. The IDers are more sophisticated in their thinking, more professional in their presentations and publications, and more politically successful in their ability to gain a public hearing for their cause. That cause, however, is the same as it has always been: to promote a Judeo-Christian biblical cosmology and worldview, to defuse any perceived threats to their religion (such as science and evolutionary theory), and to tear down the wall separating church and state in order to get their doctrines taught in public schools, including and especially public school science classes.

Chapter 12, "History's Heretics," is the oldest essay in the book, written initially while in graduate school in the late 1980s, and redacted over the years as I thought more and more about who and what mattered most in history. The germination of this project, in fact, dates back to the early

1970s when I was an undergraduate and one of my professors, Dr. Richard Hardison, introduced me to a book titled *The 100*, by Michael Hart. In the book an attempt is made to rank the top hundred people in history by their influence and importance (not just fame or infamy). Ever since then I thought about this as I studied the great (and not so great) people of the past, and when the millennium came and endless commentaries were published to venerate (or scorn) those of the past thousand years, I revised the piece yet again. Through this survey of attempts to rank the people and events most influential in our past, I devised my own ranking of who and what mattered most.

The book ends with Part IV, "Science and the Cult of Visionaries," including chapter 13, "The Hero on the Edge of Forever," an examination of Gene Roddenberry, the *Star Trek* empire, and the continuing role of the hero in science and science fiction. This is the most indulgent essay in the book because I was a fan of *Star Trek* from September 8, 1966, the date of the first airing of the first episode, a date imprinted on my psyche because it was also my twelfth birthday. But Gene Roddenberry was a humanist, which I later grew into after my stint as an evangelical Christian, and his vision of the future was far grander than any I had previously encountered. A science fiction author once explained to me that one can get away with a lot more speculation and controversy the further in the future one's narrative is set. Scientists are fairly skeptical and hard on shows like *X-Files*, because it takes place in the all-too-familiar present. But when Roddenberry put his characters into the twenty-third century, scientists were far more forgiving, and viewers glommed on to this vision of the way things could be. Still, this subject would be too self-indulgent if I had not included an analysis of what we can learn about history and social systems from an in-depth analysis of one episode of the original series—my favorite, of course—"The City on the Edge of Forever"—and what it tells us about the role of contingency and chance in our lives.

Finally, chapter 14, "This View of Science," is a comprehensive analysis of the life works of the evolutionary theorist Stephen Jay Gould, and how science, history, and biography come together into one enterprise of knowledge, understanding, and wisdom. Steve Gould was my hero, colleague, and eventually a friend, who was (and will continue to be, in my opinion) one of the most influential thinkers of our time. More than just a world-class paleontologist and world-renowned scientist, Steve was a brilliant synthesizer of data and theory—his own and others'—as well as one of the most elegant essayists of our time, perhaps of all time. Although well grounded in history, as we all should be, Gould's ideas were always on the edge of science. The primary lesson of this book is particularly well expressed in one of Gould's (and my own) favorite quotes from Charles Darwin: "How odd it is that anyone should not see that all observation must be for or against some view if it is to be of any service!"

PART I

SCIENCE
AND THE VIRTUES
OF NOT KNOWING

Psychic for a Day

Or, How I Learned Tarot Cards, Palm Reading,
Astrology, and Mediumship in Twenty-four Hours

ON WEDNESDAY, JANUARY 15, 2003, I filmed a television
show with Bill Nye in Seattle, Washington, for a new PBS science series
entitled *Eye on Nye*. This series is an adult-oriented version of Bill's wildly
successful hundred-episode children's series *Bill Nye the Science Guy*.
This thirty-minute segment was on psychics and talking to the dead.
Although I have analyzed the process and written about it extensively in
Skeptic, Scientific American, and *How We Believe,* and on www.skeptic.
com, I have had very little experience in actually *doing* psychic readings.
Bill and I thought it would be a good test of the effectiveness of the technique and the receptivity of people to it to see how well I could do it
armed with just a little knowledge.

Although the day of the taping was set weeks in advance, I did
absolutely nothing to prepare until the day before. This made me especially nervous because psychic readings are a form of acting, and good acting takes talent and practice. And I made matters even harder on myself
by convincing Bill and the producers that if we were going to do this we
should use a number of different psychic modalities, including tarot cards,
palm reading, astrology, and psychic mediumship, under the theory that
these are all "props" used to stage a psychodrama called cold reading
(reading someone "cold" without any prior knowledge). I am convinced

more than ever that cheating (getting information ahead of time on subjects) is not a necessary part of a successful reading.

I read five different people, all women that the production staff had selected and about whom I was told absolutely nothing other than the date and time of their births (in order to prepare an astrological chart). I had no contact with any of the subjects until they sat down in front of me for the taping, and there was no conversation between us until the cameras were rolling. The setting was a soundstage at KCTS, the PBS affiliate station in Seattle. Since soundstages can have a rather cold feel to them, and because the setup for a successful psychic reading is vital to generating receptivity in subjects, I instructed the production staff to set up two comfortable chairs with a small table between them, with a lace cloth covering the table and candles on and around the table, all sitting on a beautiful Persian rug. Soft colored lighting and incense provided a "spiritual" backdrop.

The Partial Facts of Cold Reading

My primary source for all of the readings was Ian Rowland's insightful and encyclopedic *The Full Facts Book of Cold Reading* (now in a third edition available at www.ian-rowland.com). There is much more to the cold reading process than I previously understood before undertaking to read this book carefully with an eye on performing rather than just analyzing. (Please keep in mind that what I'm describing here is only a small sampling from this comprehensive compendium by a professional cold reader who is arguably one of the best in the world.)

Rowland stresses the importance of the prereading setup to prime the subject into being sympathetic to the cold reading. He suggests—and I took him up on these suggestions—adopting a soft voice, a calm demeanor, and sympathetic and nonconfrontational body language: a pleasant smile, constant eye contact, head tilted to one side while listening, and facing the subject with legs together (not crossed) and arms unfolded. I opened each reading by introducing myself as Michael from Hollywood, calling myself a "psychic intuitor." I explained that my "clients" come to see me about various things that might be weighing heavy on their hearts (the heart being the preferred organ of New Age spirituality), and that as an intuitor it was my job to use my special gift of intuition. I added that everyone has this gift, but that I have improved mine through practice. I said that we would start general and then get more focused, beginning with the present, then glancing back to the past, and finally taking a glimpse of the future.

I also noted that we psychics cannot predict the future perfectly—setting up the preemptive excuse for later misses—by explaining how we look for general trends and "inclinations" (an astrological buzzword). I built on the disclaimer by adding a touch of self-effacing humor meant also to initiate a bond between us: "While it would be wonderful if I was a hundred percent accurate, you know, no one is perfect. After all, if I could psychically divine the numbers to next week's winning lottery I would keep them for myself!" Finally, I explained that there are many forms of psychic readings, including tarot cards, palm reading, astrology, and the like, and that my specialty was . . . whatever modality I was about to do with that particular subject.

Since I do not do psychic readings for a living, I do not have a deep backlog of dialogue, questions, and commentary from which to draw, so I outlined the reading into the following themata that are easy to remember (that is, these are the main subject areas that people want to talk about when they go to a psychic): love, health, money, career, travel, education, ambitions. I also added a personality component, since most people want to hear something about their inner selves. I didn't have time to memorize all the trite and trivial personality traits that psychics serve their marks, so I used the Five Factor Model of personality, also known as the "Big Five," that has an easy acronym of OCEAN: *openness to experience, conscientiousness, extroversion, agreeableness*, and *neuroticism*. Since I have been conducting personality research with my colleague Frank Sulloway (primarily through a method we pioneered of assessing the personality traits of historical personages such as Charles Darwin, Alfred Russel Wallace, and Carl Sagan through the use of expert raters), it was easy for me to riffle through the various adjectives used by psychologists to describe these five personality traits. For example: *openness to experience* (fantasy, feelings, likes to travel), *conscientiousness* (competence, order, dutifulness), *agreeableness* (tender-minded versus tough-minded), *extroversion* (gregariousness, assertiveness, excitement seeking), and *neuroticism* (anxiety, anger, depression). Since there is sound experimental research validating these traits, and I have learned through Sulloway's research how they are influenced by family dynamics and birth order, I was able to employ this knowledge to my advantage in the readings, including (with great effect) nailing the correct birth order (firstborn, middle-born, or later-born) of each of my subjects!

I began with what Rowland calls the "Rainbow Ruse" and "Fine Flattery," and what other mentalists more generally call a Barnum reading (offering something for everyone, as P.T. always did). The components of the following reading come from various sources; the particular sequential arrangement is my own. I opened my readings with this general statement:

> *You can be a very considerate person, very quick to provide for others, but there are times, if you are honest, when you recognize a selfish streak in yourself. I would say that on the whole you can be a rather quiet, self-effacing type, but when the circumstances are right, you can be quite the life of the party if the mood strikes you.*
>
> *Sometimes you are too honest about your feelings and you reveal too much of yourself. You are good at thinking things through and you like to see proof before you change your mind about anything. When you find yourself in a new situation you are very cautious until you find out what's going on, and then you begin to act with confidence.*
>
> *What I get here is that you are someone who can generally be trusted. Not a saint, not perfect, but let's just say that when it really matters this is someone who does understand the importance of being trustworthy. You know how to be a good friend.*
>
> *You are able to discipline yourself so that you seem in control to others, but actually you sometimes feel somewhat insecure. You wish you could be a little more popular and at ease in your interpersonal relationships than you are now.*
>
> *You are wise in the ways of the world, a wisdom gained through hard experience rather than book learning.*

According to Rowland—and he was spot-on with this one—the statement "You are wise in the ways of the world, a wisdom gained through hard experience rather than book learning" was flattery gold. Every one of my subjects nodded furiously in agreement, emphasizing that this statement summed them up to a tee.

After the general statement and personality assessment, I went for specific comments lifted straight from Rowland's list of high probability guesses. These include items found in the subject's home:

A box of old photographs, some in albums, most not in albums
Old medicine or medical supplies out of date
Toys, books, mementos from childhood
Jewelry from a deceased family member
Pack of cards, maybe a card missing
Electronic gizmo or gadget that no longer works
Notepad or message board with missing matching pen
Out-of-date note on fridge or near the phone
Books about a hobby no longer pursued
Out-of-date calendar
Drawer that is stuck or doesn't slide properly
Keys that you can't remember what they go to
Watch or clock that no longer works

And peculiarities about the person:

Scar on knee
Number 2 in the home address
Childhood accident involving water
Clothing never worn
Photos of loved ones in purse
Wore hair long as a child, then shorter haircut
One earring with a missing match

I added one of my own to great effect: "I see a white car." All of my subjects were able to find a meaningful connection to a white car. As I was reading this list on the flight to Seattle the morning of the reading, I was amazed to discover how many flight attendants and people around me validated them.

Finally, Rowland reminds his ersatz psychics that if the setup is done properly people are only too willing to offer information, especially if you ask the right questions. Here are a few winners:

"Tell me, are you currently in a long-term relationship, or not?"
"Are you satisfied in terms of your career, or is there a problem?"
"What is it about your health that concerns you?"

"Who is the person who has passed over that you want to try to contact today?"

While going through the Barnum reading I remembered to pepper the commentary with what Rowland calls "incidental questions," such as:

"Now why would that be?"
"Is this making sense to you?"
"Does this sound right?"
"Would you say this is along the right lines for you?"
"This is significant to you, isn't it?"
"You can connect with this, can't you?"
"So who might this refer to please?"
"What might this link to in your life?"
"What period of your life, please, might this relate to?"
"So tell me, how might this be significant to you?"
"Can you see why this might be the impression I'm getting?"

With this background, all gleaned from a single day of intense reading and note taking, I was set.

The Tarot Card Reading

My first subject was a twenty-one-year-old woman, for whom I was to do a tarot card reading. To prepare myself I bought a "Haindl Tarot Deck," created by Hermann Haindl and produced by U.S. Games Systems ($16) in Stamford, Connecticut, at the Alexandria II New Age Bookstore in Pasadena, California, and read through the little pamphlet that comes with it (itself glossed from a two-volume narrative that presumably gives an expanded explanation of each card). It is a sleek, elegantly illustrated deck, each card of which is replete with an astrological symbol, a rune sign, a Hebrew letter, I Ching symbols, and lots of mythic characters from history. For example, the Wheel of Fortune card description reads:

The wheel is set against a field of stars symbolizing the cosmos. Below, looking upward, is the Mother, the Earth. At the upper left is the Sky Father, Zeus. At the upper right is an androgynous child. The child, with its wizened face, represents humanity and our ancestors. Inside the Wheel, the mushrooms symbolize luck, the snake, rebirth, the eye, time, the dinosaur, all things lost in the turning of time. *Divinatory meanings*: Change of circumstances. Taking hold of one's life. Grabbing hold of fate. Time to take what life has given you.

For dramatic effect I added the Death card (figure 1.1) to my presentation.

The image of the boat belongs to birth as well as to death; the baby's cradle originally symbolized a boat. The trees and grass signify plants, the bones, minerals, the birds, the animal world, and the ferryman, the human world. The peacock's eye in the center signifies looking at the truth in regard to death. The bird also symbolizes the soul and the divine potential of a person. *Divinatory meanings*: The Death card rarely refers to physical death. Rather, it has to do with one's feelings about death. Psychologically, letting go. New opportunities.

At a total of seventy-eight cards there was no way I was going to memorize all the "real" meanings and symbols, so the night before I sat down with my family and read through the instruction manual and we did a reading together, going through what each of the ten cards we used is suppose to mean. My eleven-year-old daughter, Devin, then quizzed me on them until I had them down cold. (This was, I think, done not just out of Devin's desire to help her dad; it also had the distinct advantage of getting her out of doing her homework for the evening, plus gave me a taste of my own medicine of repetitive learning.) I used what is called the Hagall Spread (no explanation given as to who or what a Hagall is), where you initially lay out four cards in a diamond shape, then put three cards on top and three more on the bottom (figure 1.2). This is what the spread is suppose to indicate:

1. The general situation
2. Something you've done, or an experience you've had that has helped create the current situation
3. Your beliefs, impressions, and expectations, conscious or subconscious, of the situation

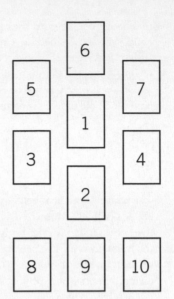

Figure 1.1 (*left*).   The Death card

Figure 1.2 (*right*).   The Hagall Spread of tarot cards

4. The likely result of the situation as things stand now
5. Spiritual history, how you've behaved, what you've learned
6. Spiritual task at this time, challenges and opportunities in the current situation
7. Metamorphosis, how the situation will change, and the spiritual tasks that will come to you as a result
8. The Helper. Visualize the actual person. This person gives you support
9. Yourself. You are expressing the qualities of the person shown on the card
10. The Teacher. This figure can indicate the demands of the situation, and also the knowledge that you can gain from the situation

By the time of the reading I forgot all of this, so I made up a story about how the center four cards represent the present, the top three cards represent the future, and the bottom three cards are the characters that are going to help you get to that future. It turns out that it doesn't matter what story you make up, as long as it sounds convincing. I was glad, however, that I had memorized the meanings of the symbols and characters on the cards I used because my subject had previously done tarot card reading herself. (Since you are supposed to have the client shuffle the tarot cards ahead of time to put her influence into the deck, I palmed my memorized cards and then put them on top of the newly shuffled deck.)

Since this subject was my first reading I was a little stiff and nervous, so I did not stray far from the standard Barnum reading, worked my way through the Big Five personality traits fairly successfully (and from which I correctly guessed that she was a middle child between firstborn and last-born siblings), and did not hazard any of the high-probability guesses. Since she was a student I figured she was indecisive about her life, so I offered lots of trite generalities that would have applied to almost anyone: "you are uncertain about your future but excited about the possibilities," "you are confident in your talents yet you still harbor some insecurities," "I see travel in your immediate future," "you strike a healthy balance between head and heart," and so forth.

Tarot cards are a great gimmick because they provide the cold reader with a prop to lean on, something to reference and point to, something for the subject to ask about. I purposely put the Death card in the spread because that one seems to make people anxious (recall the Death card was in the news of late because the East Coast sniper of 2002 said that it influenced him to begin his killing spree). This gave me an opportunity to pontificate about the meaning of life and death, that the card actually represents not physical death but metaphorical death, that transitions in life are a time of opportunity—the "death" of a career and the "rebirth" of another career—and other such dribble. The bait was set and the line cast. I had only to wait for the fish to bite.

After each reading the producers conducted a short taped interview with the subject, asking them how they thought the reading went. This young lady said she thought the reading went well, that I accurately summarized her life and personality, but that there were no surprises, nothing that struck her as startling. She had experienced psychic readings before and that mine was fairly typical. I felt that the reading was mediocre at best. I was just getting started.

The Palm Reading

My second reading was on a young woman aged nineteen. Palm reading is the best of the psychic props because, as in the tarot cards, there is something specific to reference, but it has the added advantage of making physical contact with the subject. I could not remember what all the lines on a palm are suppose to represent, so while I was memorizing the tarot cards, my daughter did a Google image search for me and downloaded a palm chart (figure 1.3).

I mainly focused on the Life, Head, Heart, and Health lines, and for added effect added some blather about the Marriage, Money, and Fate lines. Useful nonsense includes:

> If the Head and Life lines are connected, it means that there was an early dependence on family.
> If the Head and Life lines are not connected, it means the client has declared independence early.
> The degree of separation between the Head and Heart lines indicates the degree of dependence or independence between the head and the heart for making decisions.
> The strength of the Head line indicates the thinking style—intuitive or rational.
> Breaks in the Head line may mean there was a head injury, or that the subject gets headaches, or something happened to the head at some time in the subject's life.

On one Web page I downloaded some material about the angles of the thumbs to the hand that was quite useful. You have the subject rest both hands palm down on the table, and then observe whether they are relaxed or tight and whether the fingers are close together or spread apart. This purportedly indicates how uptight or relaxed the subject is, how extroverted or introverted, how confident or insecure, etc. According to one palm reader a small thumb angle "reveals that you are a person who does not rush into doing things. You are cautious and wisely observe the situation before taking action. You are not pushy about getting your way." A medium thumb angle "reveals that you do things both for yourself and for

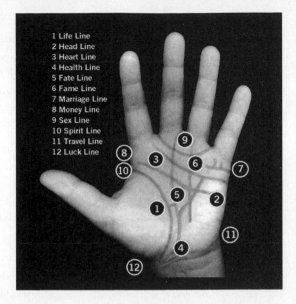

Figure 1.3.
A palm reading chart

others willingly. You are not overly mental about what you are going to do, so you don't waste a lot of time doing unnecessary planning for each job." And a big thumb angle "reveals that you are eager to jump in and get things done right away. You do things quickly, confidently, and pleasurably because you like to take charge and get the job done." Conveniently, you can successfully use any of these descriptions with anyone.

It turns out that you can tell the handedness of a person because the dominant hand is a little larger and more muscular. That gave me an opening to tell my mark, who was left-handed, that she was right-brain dominant, which means that she puts more emphasis on intuition than on intellect, that she is herself very intuitive (Rowland says that a great ruse is to flatter the subjects with praise about their own psychic powers), and that her wisdom comes more from real-world experience than traditional book learning. She nodded furiously in agreement.

According to the various palm reading "experts," you are suppose to comment on the color and texture of the skin, hair on the back of the palm, and general shape of the hands. Any major discrepancy between the two hands is supposed to be a sign of areas where subjects have departed from their inherited potential. The psychic should also take note of the shape of the fingers. The outer phalanges of the fingers (the fingertips) represent spiritual or idealistic aspects of the person, the middle phalanges everyday and practical aspects, and the lower phalanges the emotional aspects of personality. I found it most effective to rub my fingers over the mounds of flesh on each finger segment while commenting on her personality.

For this reading I threw in a few high-probability guesses, starting with the white car. It turns out that this subject's ninety-nine-year-old grand-mother had a white car, which gave me an opening to comment about the special nature of her relationship with her grandmother, which was spot-on. Then I tried the out-of-date calendar, which did not draw an affirma-tive response from my subject, so I recovered by backing off toward a more general comment: "Well . . . what I'm getting here is something about a transition from one period in your life to another," which elicited a positive affirmation that she was thinking of switching majors.

This subject's assessment of my reading was slightly more positive than the first subject's, as was my own self-evaluation, but no one was yet floored by anything said. I was still gathering steam for the big push to come.

The Astrological Reading

My third subject was another young lady, age twenty, and was my toughest read of the day. She gave monosyllabic answers to my incidental questions for extracting information and did not seem to have much going on in her life that required psychic advice. I downloaded an astrological chart from the Internet (figure 1.4). It was constructed some time ago for a guy named John, born May 9, 1961. My subject was born September 3, 1982. I haven't got a clue what the chart is supposed to mean, but since it doesn't actually mean anything I began by explaining that "the stars incline but do not compel," and then I just made up a bunch of stuff about how the con-junction of having a rising moon in the third house and a setting sun in the fifth house is an indication that she has a bright future ahead of her, and that her personality is a healthy balance between mind and spirit, intellect and intuition. Her nods indicated agreement.

I tossed out a bunch of high-probability guesses and got about half of them right (including the line about wearing her hair longer at a younger age), then closed the reading by asking her if she had any questions for me. She said that she had applied for a scholarship in order to participate

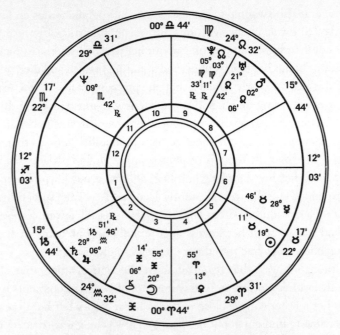

Figure 1.4. The astrological chart used by the author

in a foreign student exchange program in England and wanted to know if she was going to get it. I responded that the important issue at hand was not whether she was going to get the scholarship or not, but how she would deal with getting it or not getting it, and that I was highly confident that her well-balanced personality would allow her to handle whatever the outcome. This seemed to go over well. In the post-reading interview she was much more positive than I anticipated, considering how stilted the reading was, so I suppose we can count this one as a success as well, although I was not particularly proud of it.

Psychic Reading

My fourth subject was a fifty-eight-year-old woman, for which I was to do a straight cold read with no props. I began with the Barnum reading, but did not get far into it before it became apparent that she was more than a little willing to talk about her problems. She did not want to hear all that generic crap. She wanted to get straight to the specific issues on her mind that day. She was overweight and did not look particularly healthy, but since I didn't want to say anything about her weight I said I was picking up something about her being concerned about her health and diet, guessing, since this was still early January, that she probably made a New Year's resolution about losing weight and starting a new exercise program. Bingo!

This subject then opened up about her recent back surgery and other bodily ailments. I tried a number of high-probability guesses that worked quite well, especially the box of photographs, broken gadgets around the house, and the short hair/long hair line, all hits, especially the hair, which she explained she changes constantly. I said I was getting something about a scar or scrape on her knees, and that left her slack jawed. She said that she had not scraped her knees since childhood but had just the week before fallen down and torn them up pretty badly. Swish!

Although I was able to glean from the conversation that she had recently lost her mother, and a few minutes of generic comments from me about her mother staying close to her in her memory left her in tears, she really came to find out about her son. What was he going to do? A minute of Q & A revealed that he is a senior in high school, so I assumed that as his mother she was worried about him going off to college. Nothing but net! What in particular was she worried about? He was thinking of going to USC, so I jumped in before she could explain and surmised that it was because the University of Southern California is located in downtown Los Angeles, not exactly the safest neighborhood in the area.

In the post-reading interview this subject praised my psychic intuition to the hilt and Bill and his producers were beside themselves with glee at what great dramatic television this was going to make. Imagine how John Edward's producers must feel after a taping of *Crossing Over*. (On a positive note we did learn that day that psychic James Van Praagh's television series had just been canceled due to poor ratings.)

Talking to the Dead

My last subject was a woman age fifty, who turned out to be my best reading. She had told Bill's producer that she had something very specific she wanted to talk about, but did not offer a clue as to what it was. It didn't take me long to find out. When I introduced myself and shook her hand, I noticed that her hands were exceptionally muscular and her palms sweaty. This was a high-strung, nervous person who was obviously emotional and agitated. I assumed that someone near to her had died (the proper phrase is "passed into spirit") and that she wanted to make contact. "I'm sensing several people that have passed over, either parents or a parentlike figure to you." It was her father who died, and she clearly had unfinished business with him.

From the ensuing conversation I discovered that her father had died when she was twenty-seven, so I deduced that it must have been a sudden death and that she did not have the opportunity to make her peace with him (both correct). Finally, I accurately deduced that she was sad because she would have liked to share her many life experiences over the past two decades with her father . . . "such as having a child." Wrong—she is childless. Without missing a beat I offered this riposte: "Oh, what I mean is . . . giving birth to a new idea or new business." A three-pointer from downtown! This was an entrepreneurial woman whose father was a successful businessman with whom she would have loved to share the success.

It wasn't long before my charge was nearly sobbing. This was an emotionally fragile woman of whom I could have easily taken advantage by jumping in with some inane line such as "Your father is here with us now and he wants you to know that he loves you." But I knew I would have to look at myself in the mirror the next morning and just couldn't do it, even for a worthwhile exposé of a very evil practice. Instead I said, "Your father would want you to keep him in your heart and your memories, but it is time now to move on." I wanted to give her something specific, as well as lighten up the reading because it was getting pretty glum, so I said, "And it's okay to throw away all those boxes of his stuff that you have been keeping but want to get rid of." She burst out laughing and confessed that she had a garage full of her father's belongings that she had long wanted to dispose of but was feeling guilty about doing so. This exchange was, I hoped, a moral message that violated no trust on my part and still had the desired effect for our show.

In the post-reading interview this subject said that she had been going to psychics for over ten years trying to resolve this business with her father, and that mine was the single best reading she had ever had. Wow! That made my psychic day.

The Disclosure

Our original plan was to tell all of the subjects after the readings that this was all just a setup to expose psychic phonies as nothing more than scam artists out to rip off people. We were not particularly worried about the college-aged subjects because, after all, if the biggest crisis in your life is trying to decide whether to major in theater arts or English lit, the veracity of a psychic's reading is a minor wrinkle in the greater cosmic scheme. But for the last two readings we decided that they needed to be handled a little more delicately. Bill and the producers came up with this explanation: "We wanted you to know that Michael is not a psychic. He is a psychologist. We wanted to see if he could do what psychics do in terms of so-called intuitive readings. While we realize that this probably seemed real to you, it was not real for Michael in any psychic sense."

The Grace of Scruples

I am not a psychic and I do not believe that ESP, telepathy, clairvoyance, clairaudience, or any of the other forms of psi power has any basis whatsoever in fact. There is not a shred of evidence that any of this is real, and the fact that I could do it reasonably well with only one day of preparation shows just how vulnerable people are to these very effective nostrums. I can only imagine what I could do with more experience. Give me six hours a day of practice for a couple of months and I have no doubt that I could easily host a successful syndicated television series and increase by orders of magnitude my current bank balance—there for the grace of evolved moral sentiments and guilt-laden scruples go I. I cannot do this for one simple reason—it is wrong. I have lost both of my parents—my father suddenly of a heart attack in 1986, my mother slowly from brain cancer in 2000—and I cannot imagine anything more insulting to the dead, and more insidious to the living, than constructing a fantasy that they are hovering nearby in the psychic ether, awaiting some self-proclaimed psychic conduit to reveal to me breathtaking insights about scarred knees, broken appliances, and unfulfilled desires. This is worse than wrong. It is wanton depravity.

The Big "Bright" Brouhaha

An Empirical Study on an

Emerging Skeptical Movement

IN 1999 A GALLUP POLL inquired of Americans: "If your party nominated a generally well-qualified person for president who happened to be an X would you vote for that person?" X represents Catholic, Jew, Baptist, Mormon, black, homosexual, woman, or atheist. Although six of the eight received more than 90 percent approval—showing that America has become a more tolerant and ecumenical society—only 59 percent said they would vote for a homosexual, and less than half, 49 percent, would vote for an atheist.

Words matter and language counts. *Feminist* is a fine word that describes someone who believes in the need to secure the rights and opportunities for women equivalent to those provided for men. Unfortunately, thanks to certain conservative commentators, it has also come to be associated with sandal-wearing, tree-hugging, postmodern, deconstructionist, left-leaning liberals best scorned as "Femi-Nazis."

Likewise, *atheist* is a descriptive term that simply means "without theism" and describes someone who does not believe in God. Unfortunately, thanks to religious fundamentalists, it has also come to be associated with sandal-wearing, tree-hugging, postmodern, deconstructionist, left-leaning liberals who are immoral, pinko Communists hell-bent on corrupting the morals of America's youth.

Speak the scorn into existence.

The "Brights" Are Born

At the April 2003 conference of the Atheist Alliance International in Florida, at the behest of the organizer, I spoke about this labeling problem in the context of what "we" should call ourselves: skeptics, nonbelievers, nontheists, atheists, agnostics, heretics, infidels, free thinkers, humanists, secular humanists, and the like. Apparently there was some discussion among the organizers about whether or not I should be invited to speak because in my book *How We Believe* I defined myself as an agnostic instead of atheist, by which I mean, as Huxley originally defined the term in 1869, that the question of God's existence is an insoluble one. I suggested, rather strongly, that dividing up the skeptical and humanist communities because of a disagreement over labels was not dissimilar to the Baptists and Anabaptists squabbling (and eventually splitting) over the appropriate time for baptism (infancy vs. adulthood).

My lecture was followed by a formal PowerPoint presentation introducing a "new meme," by Paul Geisert and Mynga Futrell, from Sacramento, California, who noted that, by analogy, homosexuals used to suffer a similar problem when they were called homos, queers, fruits, fags, and fairies. Their solution was to change the label to a more neutral term—*gay*. Over the past couple of decades, gays have won significant liberties for themselves, starting with gay pride and gay marches that have led to gay rights. Analogously, instead of calling ourselves skeptics, nonbelievers, nontheists, atheists, agnostics, infidels, heretics, free thinkers, humanists, secular humanists, and the like, it was suggested that we call ourselves *brights*.

What is a bright? As defined by its creators, "A Bright is a person whose worldview is naturalistic—free of supernatural and mystical elements. Brights base their ethics and actions on a naturalistic worldview." At present there is no brick-and-mortar brights headquarters, no brights secret handshake or decoder ring. This is a cyberspace phenomenon that "seeks unification of these many persons into an Internet constituency that will grow to have significant social and political influence." Given our "severe linguistic disadvantage," the codirectors state, "The Brights movement asks those with a naturalistic worldview to join hands (in a metaphorical sense) and to begin to view themselves and speak in civic situations as

Brights." As such we must unite against the prejudice that as nonbelievers we are not qualified to be full participatory citizens. "Brights have as full a spectrum of beliefs as any other citizens. We, as Brights, reject entirely societal imposition of a set of distasteful negative labels (unbelievers, godless, irreligious) that hamper us unfairly as we work to present our views in the civic arena."

The Reaction to "Brights"

The feedback from audience members was difficult to read, but immediately following the presentation, myself, the Oxford University evolutionary biologist Richard Dawkins, the magician and skeptic James Randi, and many others all formally signed up to be brights. In the months following the conference Dawkins penned an opinion editorial for the June 21 issue of *The Guardian* in London, and the Tufts University philosopher Daniel Dennett announced in the *New York Times* on July 12 that he too was a bright. At a June conference in Seattle for gifted high school students at which both Dennett and I spoke about being nonbelievers and brights, there was an enthusiastic reception that generated a buzz among both students and speakers the rest of the weekend. It seems that lots of teenagers, and even many adult professionals from all walks of life, are nonbelievers but have been reluctant to come out as such for fear of retribution.

In his commentary, Dawkins opined that "as with gays, the more brights come out, the easier it will be for yet more brights to do so. People reluctant to use the word *atheist* might be happy to come out as a bright." Dawkins admitted that the phrase "I am bright" rings with arrogance, so he hopes that "I am *a* bright" will solicit inquiry. "It invites the question, 'What on earth is a bright?' And then you're away: 'A bright is a person whose worldview is free of supernatural and mystical elements. The ethics and actions of a bright are based on a naturalistic worldview.'" Dennett, in turn, announced, "The time has come for us brights to come out of the closet" and admit publicly that we "don't believe in ghosts or elves or the Easter Bunny—or God." He too averred: "Don't confuse the noun with the adjective: 'I'm a bright' is not a boast but a proud avowal of an inquisitive worldview."

Where Dawkins observed that "brights constitute 60% of American scientists, and a stunning 93% of those scientists good enough to be elected to the elite National Academy of Sciences" (referring to a study conducted in 1996 by Edward Larson), Dennett added that brights are "all around you: we're doctors, nurses, police officers, schoolteachers, crossing guards and men and women serving in the military. We are your sons and daughters, your brothers and sisters. Our colleges and universities teem with brights. Wanting to preserve and transmit a great culture, we even teach Sunday school and Hebrew classes. Many of the nation's clergy members are closet brights, I suspect. We are, in fact, the moral backbone of the nation: brights take their civic duties seriously precisely because they don't trust God to save humanity from its follies."

How many brights are there? Officially there are several thousand from seventy-nine nations. Unofficially, Dennett cited a 2002 survey by the Pew Forum on Religion and Public Life, which concluded that twenty-seven million Americans are atheists, agnostics, or claim no religious preference. "That figure may well be too low," Dennett suggested optimistically, "since many nonbelievers are reluctant to admit that their religious observance is more a civic or social duty than a religious one—more a matter of protective coloration than conviction."

Reactions from public intellectuals came swiftly. A number of people reported that they heard Rush Limbaugh comment on his syndicated radio show heard by tens of millions of people that brights think they are "brighter than those who believe in God." In *The Times* of London Ben McIntyre confessed on July 19, 2002, "I shall not be coming out as a Bright just yet. For a start, the term 'secular humanist' may be old-fashioned but it is still serviceable, and mercifully doesn't sound like something dreamed up as an advertising gimmick." He noted, with classic British humor, "It has the added advantage that the Religious Right in America already loathes it, so it must be just fine." More seriously, McIntyre observed, "The term 'Bright' seems too all-embracing for so many shades of doubt and certainty. Why, in rejecting the extravagant claims of organised religion, would one want to be part of a group organised around the absence of religion? Almost by definition, Brights would be opposed to joining any club clamouring to have them as a member, and for that reason alone I suspect the Brights may dim pretty quickly."

Even National Public Radio—one of the last liberal bastions left on the airwaves and long supportive of the ACLU and other secular causes—

aired a critical commentary. On September 4, 2002, NPR's Steven Wald-man asked, "Are atheists and agnostics smarter than everyone else?" and answered with the already oft-quoted antonym: "I'm not sure what the image buffers were aiming for, but the name 'The Brights' succinctly conveys the sense that this group thinks it's more intelligent than everyone else. The rest of us would be 'The Dims,' I suppose." Waldman admits, "Our political culture has increasingly marginalized atheists, agnostics, and secular humanists—whether it's ten commandments in the courtroom, or a president who invokes god continuously," and that "any rationalist would be perfectly justified in thinking society views them as second class citizens whose views are not worthy." Indeed. But then he sets up a straw-man argument by claiming that brights have asserted "that people who believe in god or the supernatural are just not as, well, bright." To my knowledge no bright has made such an assertion, and if they did they would be wrong since survey data show (as Waldman notes) that over half of all Americans with postgraduate degrees believe in the devil, Hell, miracles, the afterlife, the virgin birth, and the resurrection. As I have demonstrated elsewhere (*Why People Believe Weird Things*), smart people believe weird things because they are better at rationalizing beliefs they arrived at for nonsmart (psychological and emotional) reasons.

Dinesh D'Souza, in an October 6, 2002, *Wall Street Journal* editorial, began by noting that our movement claims a heritage dating back to the Enlightenment, then argued that our fundamental belief in the power of science and reason to encompass all knowledge about the world was thoroughly debunked by the German Enlightenment philosopher Immanuel Kant, who demonstrated that no system of human knowledge could be complete, and thus there will always be epistemological limits beyond which such entities as God might exist. That is, of course, possible, and no scientist or skeptic ever claimed that science is omnipotent or omniscient. But the possibility that the divine may exist in some realm beyond the reaches of science by no means proves that it does, and D'Souza offers no enlightenment, as it were, on this front.

For my part I wrote a new ending to my book, *The Science of Good and Evil*, and an opinion editorial based on that section, in which I said: "*Bright* is a good word. It means 'cheerful and lively,' 'showing an ability to think, learn, or respond quickly,' and 'reflecting or giving off strong light.' Brights are cheerful thinkers who reflect the light of liberty and tolerance for all, both brights and non-brights. It is time for brights, like other minorities in our country, to stand up and be counted, either literally at www.thebrights.net, or figuratively the next time a politician kowtows to the religious right with a gratuitous slam against those of us who do not believe in God or are nonreligious. Just as it is no longer okay to woman-bash or gay-bash, it is now unacceptable to bright-bash. Saying so may not make it so, but all social movements begin with the word. The word is bright."

Study 1: Unsolicited Feedback on "Brights"

At the behest of my publisher I decided to hold off submitting my opinion editorial until my book *The Science of Good and Evil* arrived in stores in February 2004. Instead, I "came out" to the 25,000 readers of our electronic *e-Skeptic* newsletter. To my astonishment I received hundreds of e-mails, the vast majority of which were emphatically negative about the term. Most strongly indicated that under no circumstances would they call themselves brights. The primary reason given was that the word sounds elitist, especially since the natural antonym is "dims." One correspondent, Joseph Giandalone from Conway, Massachusetts, wrote:

> I'm a 55-year-old scientist/humanist/atheist since my early twenties and I've thought about these things for many years and I am pained to tell you that your choice of the term "Bright" as the one to promote is a horrible one. I agree entirely and enthusiastically with your enterprise and the reasoning that goes into it, but I am dumbfounded that you would choose a term that will do nothing more than expose us to ridicule and engender hostility in those who do not agree with our worldview. Consider two facts: (1) In the popular lexicon, "bright" as applies to people means "smart." (2) Believers in God (and etc.) *really resent us already* because we have the gall to reject their most cherished beliefs and to imply that people like them must be morons if they believe as they do. Put 1 and 2 together, please!

Since I had not kept track of how many positive and negative responses were received, and because I had not even invited feedback and commentary, I decided to initiate a slightly more formal study by posting the above letter in a second *e-Skeptic*, this time asking for feedback and commentary from readers. We received many thoughtful letters, such as this one from Ralph Leighton, the coauthor, with the Nobel laureate Richard Feynman, of the book *Surely You're Joking, Mr. Feynman*:

> "Rationalist" is the closest word to describe my beliefs. What's missing is the deep emotions that one can get from a rational belief. In fact, the awe that one feels for the universe is deeper, in my view, when one looks at the universe rationally, than when one views it irrationally, i.e., as a folk belief or religious story. That's why I came up with the term "Passionate Rationalist" to describe this seeming contradiction—but I concede that it is too long an expression to be practical. By the way, when it comes to whether there is a god, I think I remember Feynman describing himself as a "non-believer." When I asked him what he meant, he said, "You describe it; I don't believe in it." Feynman was not saying he didn't believe there was a god; he was saying that any god that you can describe is too limiting for him to believe in. Michael, I strongly agree with the writer that there is a danger of setting back the cause. I wouldn't just let it go and see if it sticks. I would try to retract the term ASAP.

The actress (*Clock Stoppers*) and comedian (*Saturday Night Live*) Julia Sweeney, whose one-woman show *Letting Go of God* chronicles her journey from Catholicism to skepticism, also disliked the brights label, explaining that it sounds nerdy and unsexy. "When people ask, I say I'm a cultural Catholic and a philosophical Naturalist, natural being the opposite of supernatural."

Study 2: Solicited Feedback on "Brights"

In all we received eighty-nine letters. Eight were clearly positive; the rest ranged from neutral to highly critical. Six of the eight positive letters included commentary, all of which are included below. Of the remaining letters, I include excerpts from twenty-two of them below.

Positive Reactions to "Bright"

> First of all, I think trying to solicit alternatives to the word "bright" is an exercise in futility. There will always be a small vocal minority who can't stand any particular word for one reason or another. The

point, oh clueless ones, is not the stupid word. The point is that together, we (being those of us with a naturalistic worldview) can be a force to be reckoned with. The point is, dear fellow atheists/skeptics/humanist/free-thinkers/hummingbirds, the Canadian Parliament is not going to take, say, 500 Canadian Humanists seriously (who can't even get their shit together among all the individual chapters). But add up all the Canadian humanist groups, skeptic groups, and all those folks who checked "no religious affiliation" on the last census, and you've got at least a million Canadian Brights. In Canada, that's a lot of votes.

I like the term "Bright." I think it is interesting and proper and those arguments against it are unfounded and based on fear of what others (like believers) think. Cowards. The term "Bright" will stick. It should. It sounds good and I liked it since I had first seen it associated with James Randi. I didn't know what it meant at first, until you sent me your *e-Skeptic,* but now that I know, my instincts of it were confirmed and the term fits Randi and others who I think are truly Brights. This term is distinguished and even curious for those who might not know what it means at first, but can only speculate until they find out. But one thing is for sure, they will know they aren't Brights if they aren't in the category of James Randi and yourself and others who know how to debunk their collective jargon with good science. Stick with the term, for I think it will stick. It is naturalist and free of metaphysical claims and jargon, but full of just plain and simple commonsense reality. Be well and be a Bright.

I think that "Bright" is quite neutral compared to some of the other options. As far as the argument that bright makes us sound like we think we are smarter than them, so what? I'm sorry, and I certainly don't mean to suggest that atheists are necessarily smarter than theists, but the resentment that a person aims at others who challenge his sacred beliefs will not be mollified because the challenger calls himself a "humble" or "apologist."

I loved the term when I first read about it. I felt a bit of joy, thinking that I could take that label for myself, my worldview, and share it with other individuals. "Hell yeah!" I thought.

I would like to weigh in with a tepidly positive vote. In fact I went ahead and signed up as a Bright. The point of the whole exercise is not to go around proclaiming that we are Brights and doing battle with evangelicals at every corner. My interpretation is that if we atheists, agnostics, skeptics, etc., want to join together as a more solid and cohesive voice, we must do so under one banner. I know that our ilk primarily keep their opinions to themselves and are a more solitary breed. Unfortunately, if we are to have any kind of say in what is going on in our communities and the world, then we really must come together as one informed voice and presence. If the best way to begin this process is to take on the name "Bright," then so be it.

Anyone who has endorsed the meme is not likely to write to say "me too," or "right on bro." I know that the idea has stimulated a lot of lively discussion in humanist/atheist/free-thought fora and that there are a lot of people who are downright threatened by it. Your involvement and that of Randi, Dawkins, Dennett and others, in print and on the radio, have really helped to get the idea about Bright into the open. In addition, it has made people realize that nontheists are not content to be negatively defined (or defiled) by believers. Even negative feedback spreads the meme. I don't know how successful the Bright constituency idea will be in the long run, but I think that the discussion it has stimulated both within free-thought·community, but also in the main stream, has been a good thing.

Negative Reactions to "Bright"

The response to the "bright" meme proposal immediately put me in mind of what happened here in Canada with the yearly conference of social scientists known as The Learned Societies Conference. The conference was colloquially known as The Learneds, but the society has recently changed after a backlash in many host cities had the locals referring to the conference attendees as The Stupids.

Let me add my voice to the myriad of others who have expressed opposition to any attempt to promote "Bright" as a popular euphemism for "nontheist." While I would be delighted to be publicly proclaimed as a Bright, there is not a snowflake's chance on the sunny side of Mercury of the ignorant godworshippers (tautology) ever accepting a word commonly understood to mean brainy as a description of persons who disagree with them.

The name "Brights" sounds too much like happy shiny people—sort of hippie-dippie, brain dead. And just like "child-free" in the '70s, it sounds smug. What's wrong with unbeliever? You can't be a bigger atheist than I am, yet I'm thrilled to tiny bits to handle money that says "In God We Trust" and I cheerfully sing Christmas carols. I just don't believe in anything. I'm walking, talking electric meat.

As a skeptic myself, I find this name "brights" ludicrously conclusionary and almost pitifully self-aggrandizing to the point where it calls into question—in my mind—your basic common sense. I suggest you try to swallow your pride, admit your error and drop the thing ASAP if not sooner.

Why don't we simply call ourselves "smart-people-unlike-you-dumb-people"? How to win friends and influence people.

Before I saw your piece in *e-Skeptic*, I had already signed up and sent this to the "Brights" themselves: "I signed up yesterday, because I believe fully in what you are doing. However, you *have* to come up with a better name for the movement. I did a quick Web search, and already Rush Limbaugh is vilifying the group as thinking they are "brighter than those who believe in God." This makes it difficult for me to admit to my family that I'm part of this movement, since they listen to Limbaugh and will immediately take it as some sort of superior statement, which doesn't seem at all your intention."

In the '50s my aunt was already one of the highest ranking women in the State Department—a time and a place where women in authority were rare and "ungodly Communists" were persecuted. When asked about religion, she would simply say, "I am of a scientific mind."

Let's just keep this plain and simple and out front: *atheist*. This bright stuff is just dumb. We shouldn't let others change our name because *they* have slandered it. *Atheist* is not a bad word, we should not be ashamed to use it.

Although nothing is certain, some things come close. The sun will rise and set tomorrow. An apple will fall from a tree. A birth will occur. And "bright" will remain a terrible name for us non-believers.

Its in-your-face connotation is obvious to me and everyone I talk to about it. Surely you are bright enough to see that.

Actually, I am liking the term *skeptic* more and more as time goes by. It doesn't seem to evoke the anti-christian connotations you get when you use *atheist*. Maybe it is because it seems more scientific, as opposed to relating to religion. You can be skeptical about many things, but you can only be atheist in regard to religion.

From the very first time I heard this "bright" idea a couple of months ago I thought it was dumb. After thinking about it some more, the conclusion I've come up with is that it's even dumber than I originally thought.

Register another vote against "Bright," mostly for the reason cited by the reader-response you included. The retiree's association of the company I worked for (TRW) could not agree upon a name for their newsletter. So, for over twenty years it has been known as *The No-Name Gazette*. I respectfully submit that we are better served by No-Name than by "Bright."

In the May 2003 *Atlantic Monthly*, Jonathan Rauch suggested apatheism. "The modern flowering of 'apatheism'—a disinclination to care all that much about one's own religion, and an even stronger disinclination to care about other people's—is worth getting excited about, particularly in ostensibly pious America." It isn't a synonym for the all-encompassing term "bright" but, due to a summer backlog in reading, I encountered apatheism and brights on the same day. Of the two, I like apatheism since it expresses my own views on religion very well—don't bug me with your religion and I won't bug you with mine. On the other hand, I don't much like "brights." It requires explanation, and since I'm an apatheist I don't really care to get into explanations.

I can't believe you folks are this out of touch. You are, despite your worthy intentions, doing all of us a great disservice and can only wind up setting our cause back, which we do not need. I find the fact that a number of you have decided to label People Like Me "The Brights" to be *embarrassing*. I haven't thought of a better term to use, but there have got to be many. Can't you instigate some kind of retraction and make an effort to get some kind of input from a large number of us? Get a larger sampling of opinion on this???! It's too good an idea to screw up with that horrendous choice of a label.

For God's sake (and for Gould's!) please don't call us "Brights."

Just wanted to throw in my 2-cents on the suggested rechristening of skeptics as "Brights." Bad. Bad. Goofy. And bad.

First, I don't think that re-branding is a solution. There are already a plethora of accurate terms in popular use (atheist, humanist, secularist . . .) and anyone sympathetically inclined won't have any trouble wearing one. However, anyone opposed certainly won't be placated by a cuter, cuddlier name. These are people that curse words like "science, tolerance, liberal, and establishment clause." Many of them look down on the term "lovers." We're not going to get anywhere looking for a new label. Second, it would be hard to come up with a less hip, goofier name than "Bright." It's kind of like calling your band, "The Sunshines" (unless it was death metal and the irony palpable). Ever know a wimp who changed his name to "Bruce"? Or a plain-jane rebranding herself "Roxanne"? Could you look at them without thinking, "Ridiculous?" And those are both cooler names than "bright."

I agree 100% with Mr. Giandalone's letter. Life is hard enough for atheists these days without some wise guy—er, I mean, Bright guy—deciding that all of us should be labeled as Bright. I am insulted and outraged that somebody else should decide to label me as such.

Please please, please do not set this trial balloon afloat. Rather, *please* let the hot air out of the balloon—pronto. We don't need yet another label. We need to find pro-atheistic arguments that inspire people, which instill hope, which motivate people to be good and productive members of society.

I was intrigued by your search for a word to describe me and I looked at the Bright's Web site. After giving it some thought I reached the same conclusion as Mr. Giandalone, although I wouldn't express myself so robustly. The trouble with "bright" is that it is self-congratulatory and carries a strong flavor of superiority; I can't imagine myself using it without embarrassment. I really don't see much wrong with "agnostic." It has religious connotations but the meaning of a word can change and broaden and perhaps it's easier to retrain a horse than breed a new one.

The term "bright" (in my humble opinion) is like a "kick me" sign on my back.

I'm so glad someone (finally) asked! I don't really like the word "Bright" being foisted upon me and I would much prefer something less offensive such as "freethinker." True, it is sometimes possible to believe in God and still be a freethinker but it's still a better word than "bright," and it's popular. I conducted my own poll. I deliberately put the terms in the plural (brights, freethinkers, secularists, etc.) in order to give the term Bright a fair shot. It's at: http://ankrumschultz.tripod.com/. As you can see, Freethinkers wins for 1st, 2nd *and* 3rd choice, which shows that even people who would not make it their first choice still like it. Meanwhile, "Secularists" looks fairly strong for second place and would be a good alternative to those who think "freethinkers" is too broad. Also, although 12% of people in the poll are not Freethinkers (or Brights, etc.), Freethinker ranked among one of the least offensive terms at 2%, Secularists even better at 1% (all of them were deemed offensive by someone), while 8% thought "Brights" was offensive, the second highest score for offensiveness just behind "Enlightened Ones."

Admittedly, there is a built-in bias in this study since people are more likely to respond when they are disgruntled. To their credit, Paul and Mynga sent me all of the comments people had submitted when they signed up. Of the sixty-four presented to me, seventeen made specific reference to the potentially offensive nature of the word *brights*. That is a hefty 27 percent, a number that strikes me as rather high coming from those most positive about the concept.

I had originally suggested to Paul and Mynga that we solicit feedback from various sources before settling on a new label, but they convinced me that sometimes social movements are best driven not by committee and excessive discussion (freethinkers, humanists, and skeptics have been talking about the labeling problem for decades) but by simply moving forward with the goal of making it happen by momentum, will, and force of personality. Since much of what I do gels with this philosophy, I was initially receptive.

But then the associate director of the Skeptics Society, Matt Cooper, pointed out (based on his experience as a marketing consultant and political activist) that it is not the philosophy of the movement under debate, but the brand name. This is a branding issue, not an ideology issue. And the scientific approach to branding is to conduct focus groups and market tests to see what works. Unfortunately, this was never done for the bright brand, and as a consequence we are now embroiled in a big bright brouhaha. Thus, Matt and I analyzed all of the e-mails we received in response to the second *e-Skeptic* that solicited feedback, and followed that up with a focus group study. Table 1.1 presents the attitudes of the eighty-nine *e-Skeptic* respondents.

Suggested alternatives to "Brights" numbered 124 and are presented in table 1.2. As a quick perusal will show, compared to most of these "Brights" is a vastly superior label.

Study 3: Focus Group Feedback on "Brights"

On September 7, following the Skeptics Society Distinguished Science Lecture Series at Caltech, we assembled a focus group of thirteen first-time lecture attendees. Without knowing the purpose of the focus group, the volunteers were asked to describe the audience they had just been a part of. Their suggestions are presented in table 1.3.

TABLE 1.1

Attitudes Toward the Label "Brights" by 89 *e-Skeptic* Respondents

	number of e-mails	%
1. No position, just commented on process	3	3
2. Offered alternative name without commenting on "Brights"	23	26
3. Negative on "Brights"; offered alternative name	37	42
4. Negative on "Brights"; no alternative offered	18	20
5. Positive on "Brights"	8	9
TOTAL	89	100

TABLE 1.2

Suggested Alternatives to "Brights"

Agnahumans	Freethinkers	Naturalists	Reasonists
Agnamen	Geians	Naturals	Reductionists
Agnascepts	Godless	Naturies	Rethinkers
Agnastics	Bolshaviks	Naturists	Scepnastics
Agnostics	Gouldists	Neoclearists	Scepnostics
Anaxagorians	Heathen	Neo-gnostics	Sceptmen
Anti-theist	Huhhers	Neo-thinker	Sciencians
Apatheism	Humanists	No label at all	Scientific
Apatheists	I am of a	No-Names	Secularists
Asupernaturalists	scientific	Non-believers	Scientists
Atheists	mind	Nuffists	Secular
Athnasceps	Illuminaries	Open thinkers	Humanists
Athnastics	Infidels	Openminders	Secularists
Atomists	Inquirers	Opens	Seculars
Atoms	Inquisitors	ORBs	Seekers
B.R.I.G.H.T.s	Intellectual-	Passionate	Skepnastics
Brites	liberal	Rationalists	Skeptics
Cleariats	Intellectuals	Philosophical	Skeptmen
Clears	Lifelong	Naturalists	Smart-people-
Critical thinkers	Learners	Phrontisteries	unlike-you-
Enlightened ones	Lights	Pragmatic realists	dumb-people
Enoughists	Lucids	Probing minds	Sprites
Enrealders	Mortals	QEDs	The Happies
Epicurian	Nagnoscepts	Questioneer	Thinkers
Naturalists	Nagnoskepts	RASPs	Thinkstirrers
Epicurians	Natagnostics	Rationalists	Truists
Evaluators	Natanostics	Rational	Truth Seekers
Evaluites	Natnostics	Materialists	Twains
Evolvers	Nats	RAVENs	Unbelievers
Fallibatheists	Natural	Realders	Untheists
Forthrights	Philosopher	Realists	Wonder
Freedoubters	Naturalies	Realitivists	Worlders
Freedoubts	Naturalismists	Reasonalists	Worldlings
Frees			

TABLE 1.3

Suggested Descriptive Terms for Attendees
at the Skeptics Society Caltech Lecture

Acceptors	Defensive	Intellectual	Open thinkers
Affirmer	Eclectic	Investigators	Pioneers
Bright	Inquisitive	Lost explorers	Seers
Courageous	futurists	Nerd	
Curious			

The focus group was then asked for words that might describe all such-minded people worldwide. They offered iconoclasts, rationalists, nonbelievers, open-minded, determinists, and skeptics.

The focus group was then given a list of names culled from the 124 provided by our *e-Skeptic* correspondents. They were asked which names they liked most, which they liked least, and which would be acceptable to them. The results are in table 1.4.

The most polarizing name was Freethinkers. It was expected to be offensive or embarrassing by 38 percent of the focus group, yet was chosen as a favorite by the remaining 62 percent. Table 1.5 presents the same data in a different breakdown.

The focus group session ended with a general summary of the brights controversy. The room was polled as to how many thought the public would be offended by the label "Brights." Nine of the thirteen raised their hand. This was followed by a moderated discussion regarding the reactions of the participants to this and other suggested names. Some of their thoughts included:

- Skeptic is perceived as not believing anything, or as cynicism.
- Skeptic is all-encompassing, not just God-related—*bright* is too limiting.
- They hate us anyway. Why not piss 'em off with "Brights"?
- Naturalist brings to mind nudists or biologists.
- Atheist is perceived as very negative, agnostic as wishy-washy, undecided.
- Likes "Freethinkers": we're not saying we have answers, but neither do you.

TABLE 1.4

Alternative Names Liked Most, Least,
and In Between by Focus Group

	Most	Between	Least		Most	Between	Least
Atomists	8%	15%	23%	Rationalists	23%	46%	23%
Brights	0%	0%	69%	Realists	8%	15%	38%
Conservers	0%	0%	54%	Reals	0%	15%	38%
Critical				Reasonists	8%	23%	46%
thinkers	54%	69%	0%	Recorders	0%	0%	46%
Diogenes	0%	8%	31%	Scientific			
Fact watchers	0%	8%	38%	secularists	0%	15%	54%
Forthrights	0%	0%	54%	Scientifically			
Freedoubts	8%	8%	31%	minded	8%	31%	38%
Frees	0%	0%	46%	Scientists	8%	46%	23%
Freethinkers	62%	62%	38%	Scribes	15%	15%	46%
Humanists	0%	15%	46%	Secular			
Illuminaries	15%	15%	54%	humanists	0%	8%	69%
Inquirers	46%	54%	23%	Secularists	0%	0%	62%
Lights	0%	0%	62%	Seculars	0%	8%	54%
Lucids	8%	15%	38%	Seekers	23%	31%	23%
Naturals	0%	8%	38%	Skeptics	31%	54%	15%
Observers	23%	38%	15%	Thinkers	23%	31%	38%
Observers of				Truists	0%	0%	46%
truth	0%	0%	62%	Truth seekers	8%	15%	31%
Open minders	31%	38%	31%	Unbelievers	0%	0%	77%
Opens	0%	0%	69%	Watchers	15%	23%	23%
Pragmatic				Wonders	0%	0%	31%
realists	8%	15%	31%	Worlders	0%	0%	69%
Questioners							
(eers)	15%	31%	38%				

- Likes "Freethinkers": not threatening, aggressive. Skeptic sounds confrontational.
- Freethinker sounds like an unstructured thinker.
- How about open or open thinkers? As in open to new ideas.
- Opponents will cause any chosen name to become a negative.
- Words such as *gay* and *big bang* were coined by opponents and embraced by advocates.

TABLE 1.5

Focus Group Rankings of Alternative Names for "Brights"

	Rank	Name	Cited by
Top 5 most offensive names	1	Unbelievers	77%
	2 (tie)	Brights	69%
	2 (tie)	Opens	69%
	2 (tie)	Secular humanists	69%
	2 (tie)	Worlders	69%
Top 9 least offensive names	1	Critical thinkers	0%
	2 (tie)	Skeptics	15%
	2 (tie)	Observers	15%
	4 (tie)	Inquirers	23%
	4 (tie)	Rationalists	23%
	4 (tie)	Scientists	23%
	4 (tie)	Seekers	23%
	4 (tie)	Watchers	23%
	4 (tie)	Atomists	23%
Top 9 most favored names	1	Freethinkers	62%
	2	Critical thinkers	54%
	3	Inquirers	46%
	4 (tie)	Skeptics	31%
	4 (tie)	Open minders	31%
	6 (tie)	Rationalists	23%
	6 (tie)	Observers	23%
	6 (tie)	Seekers	23%
	6 (tie)	Thinkers	23%
Top 12 most acceptable names	1	Critical thinkers	69%
	2	Freethinkers	62%
	3 (tie)	Inquirers	54%
	3 (tie)	Skeptics	54%
	5 (tie)	Rationalists	46%
	5 (tie)	Scientists	46%
	7 (tie)	Open minders	38%
	7 (tie)	Observers	38%
	9 (tie)	Seekers	31%
	9 (tie)	Thinkers	31%
	9 (tie)	Questioners	31%
	9 (tie)	Scientifically minded	31%

What Is in a Name?

For the most part I avoid labels altogether and simply prefer to say what it is that I believe or do not believe. However, at some point labels are unavoidable (most likely due to the fact that the brain is wired to pigeon-hole objects into linguistic categories), and thus one is forced to use identity language. Whether the bright meme succeeds or not, I commend Paul and Mynga for their courage and conviction and if, in a decade or two, the brights label has the same level of social acceptance as the gays label, we will all be better off for it. What is in a name? A lot. Although I like the top-down strategy of attempting to impose some order on this linguistic chaos, in the end I don't think it can be done by fiat. Instead, I suspect that a type of Darwinian selection will drive the most natural name into general acceptance. Or perhaps diverse linguistic species will peacefully coexist within their own niches.

Until then, since the name of the magazine I cofounded is *Skeptic*, and my monthly column in *Scientific American* is entitled "Skeptic," I shall continue to call myself a *skeptic*, from the Greek *skeptikos*, or "thoughtful." Etymologically, in fact, its Latin derivative is *scepticus*, for "inquiring" or "reflective." Further variations in the ancient Greek include "watchman" and "mark to aim at." Hence, skepticism is thoughtful and reflective inquiry. Skeptics are the watchmen of reasoning errors, aiming to expose bad ideas.

Perhaps the closest fit for skeptic is "a seeker after truth; an inquirer who has not yet arrived at definite convictions." Skepticism is not "seek and ye shall find"—a classic case of what is called the confirmation bias in cognitive psychology—but "seek and keep an open mind." What does it mean to have an open mind? It is to find the essential balance between orthodoxy and heresy, between a total commitment to the status quo and the blind pursuit of new ideas, between being open-minded enough to accept radical new ideas and being so open-minded that your brains fall out. The virtue of skepticism is in finding that balance. *Skeptic* is a virtuous word.

Heresies of Science

Why You Should Be Skeptical of What
You Always Thought Was True

HUMAN BEINGS ARE PATTERN-SEEKING, storytelling animals. For tens of thousands of years we have been telling stories about the meaningful patterns we find in our world and in our lives, and for the last five thousand years we have been writing down our stories. About four centuries ago, however, something big happened that profoundly changed the process of pattern-seeking and storytelling. That something was science. For the first time in human history there arose a set of methods by which it could be determined if a pattern is real and if a story is true. Instead of just retelling stories over and over, it was now possible to refine the story to more closely match reality. Consider the following remarkable scientific story.

On October 6, 1923, the universe suddenly and without warning exploded, expanding in size from thousands to millions of light-years across. The universe's actual big bang expansion happened billions of years ago, of course, but this is the day that the astronomer Edwin Hubble first realized that the fuzzy patches he was seeing in his telescope atop Mount Wilson in southern California were not "nebulae" *within* the Milky Way galaxy but were, in fact, separate galaxies, and that the universe is bigger than anyone imagined . . . a lot bigger. We are not merely a grain of sand among a hundred billion grains on a single beach; there are, in fact, hundreds of billions of beaches, each one of which contains hundreds of billions of grains. For championing a heresy that turned out right, Hubble had the first space telescope named after him and thus achieved a form of scientific immortality.

The history of science is filled with such ego-shattering discoveries. Thanks to science, change happens so fast that the world of today is as different as when I was born, during the Eisenhower presidency, as Eisenhower's world was from George Washington's. Isn't there anything in science we can count on to be "true" in some provisional sense? Yes, of course there is. The universe constitutes all that there is. Time travel is impossible. Evolution is progressive and leads to more complex and intelligent life. Oil is a finite fossil fuel of which we are about to run out. You cannot catch cancer. The brain and spinal cord cannot regenerate.

Of these six scientific facts we can be certain, no? No. There are no "facts" in science, in the sense of something being proven 100 percent. Of course, we can make provisional conclusions about which we can be exceptionally confident—and of these we might say they are "facts"—but, strictly speaking, we must be open to new claims, and listen to the scientific heretics who challenge our most cherished assumptions, because if there is one thing that is certain in the history of science, it is that nothing is certain in science. So, as an exercise in skepticism—in the original Greek meaning of that word as "thoughtful inquiry"—let's consider some serious challenges to these deeply rooted scientific assumptions, provocatively presented as heresies:

Heresy 1: The Universe Is Not All There Is

Heresy 2: Time Travel Is Possible

Heresy 3: Evolution Is Not Progressive

Heresy 4: Oil Is Not a Fossil Fuel

Heresy 5: Cancer Is an Infectious Disease

Heresy 6: The Brain and Spinal Cord Can Regenerate

With each heresy we will consider the belief it is challenging, the alternative it offers, and the likelihood that it is correct.

The Fuzzy Factor

Since these heretical ideas differ in content and certainty, I have devised a "fuzzy factor" rating for the likelihood of them being correct. I originally developed this system for my book *The Borderlands of Science* to assess claims that fall in that fuzzy gray area between clear science and obvious pseudoscience. These borderlands beliefs may be true, but we don't know yet, so I assigned them probability figures ranging from .1 (least likely to be true) to .9 (most likely to be true). Nothing gets a zero or a 1 because in science we cannot be that certain.

Consider the color of the sky. Traditional binary logic demands that it must be either blue or nonblue, but not both. In a system of fuzzy logic, however, a fuzzy fraction is a more accurate description. At dawn on the sunrise horizon the sky might be .1 blue and .9 nonblue (or, say, .9 orange). At noon overhead the sky might be .9 blue. At dusk on the sunset horizon the sky might be .2 blue and .8 nonblue (or .8 orange). This system gives us a lot more flexibility in evaluating controversial claims, and allows us to be more tolerant of heretics, so to each of these heresies I have ascribed a "fuzzy factor" between .1 and .9, from least likely to most likely to be true. The evaluations are my own and are thus necessarily personal, so you should, in the best sense of the word, be skeptical not only of the original theories and the heretical challenges to them, but of my assessment as well. In science, as in life, it is best to check things out yourself.

Heresy 1. The Universe Is Not All There Is

In the most watched television documentary series of all time, *Cosmos*, the astronomer Carl Sagan dramatically defined his subject this way: "The cosmos is all that is, or ever was, or ever will be." That was in 1980. A lot has changed since then, including the possibility that the universe may not be all there is. There may be more . . . a lot more. The single bubble universe in which we reside, that was born in a big bang and will most likely expand forever and die with a whimper, may be only one of many, perhaps an infinite number of bubble universes, all with slightly different configurations and laws of nature. In this *multiverse*, those universes that are like ours will probably give rise to life, perhaps even intelligent life smart enough to realize that it lives in a multiverse. How deliciously recursive!

There are several heretics who champion the multiverse. The maverick Oxford University physicist David Deutsch applies the "many worlds" interpretation of quantum mechanics to the cosmos, where he argues that there are an infinite number of universes in which every possible outcome of every possible choice that has ever been available, or will be available, has happened in one of those universes. This idea is based on a spooky experiment with light that is passed through two slits and forms an interference pattern of waves on a back surface (like throwing two stones in a pond and watching the concentric wave patterns interact, with crests and troughs adding and subtracting from one another). The eerie part is when you send single photons of light through the slits they still form an interference wave pattern even though they are not interacting with other photons. One explanation for this oddity is that the photons are interacting with photons in other universes!

In this theory of "parallel universes" you could actually meet yourself and, depending on which universe you entered, your parallel self would be fairly similar or dissimilar to you. In an early episode of *Star Trek* (the original series) Mr. Spock met his doppelgänger, who sported a goatee and exhibited a rather truculent temperament. In another *Star Trek* episode (the next generation), the *Enterprise* and her crew arrived at a quantum junction at which they experienced a multitude of other *Enterprises* and crews, all having made slightly different choices with their correspondingly different outcomes (one was destroyed by the Borg, another defeated the Borg, etc.).

A different form of multiverse is suggested by the Astronomer Royal of Great Britain, Sir Martin Rees. The expansion of our bubble universe may be just one "episode" of the bubble's eventual collapse and reexpansion in an eternal cycle. Or, an "eternally inflating multiverse" may sprout other bubble universes out of collapsing black holes, all with slightly different configurations. We know that our universe can sustain life because, well, here we are! But our universe could be impoverished compared to others that might be teeming with life. This multiverse model is used by the American cosmologist Lee Smolin, who adds an evolutionary component to it that resembles Darwinian "natural selection." Like its biological counterpart, Smolin thinks that there might be a selection from different "species" of universes, each containing different laws of nature. Universes like ours will have lots of stars, which means they will have lots of black holes that collapse and create new baby universes similar to our own. By contrast, universes without stars cannot have black holes, and thus will not hatch any baby universes, and thus will go extinct. The result would be a preponderance of universes like ours, so we should not be surprised to find ourselves in a universe suitable for life. There may very well be many such biophilic universes.

Much of this speculative cosmology is in the realm of what I call borderlands science because it has yet to be tested. But that does not mean that it *cannot* be tested. The theory that new universes can emerge from collapsing black holes may be illuminated through additional knowledge about the properties of black holes. Other bubble universes might be detected in the subtle temperature variations of the cosmic microwave background radiation left over from the big bang of our own universe, and NASA recently launched a spacecraft constructed to study this radiation. Another way to test these theories might be through the new Laser Interferometer Gravitational Wave Observatory (LIGO), which will be able to detect exceptionally faint gravitational waves. If there are other universes, perhaps ripples in gravitational waves will signal their presence. Maybe gravity is such a relatively weak force (compared to electromagnetism and the nuclear forces) because some of it "leaks" out to other universes. Maybe.

Because these cosmological heresies are still in their borderland stages we must be cautious in our evaluation of them. The mathematics may be "beautiful" and the theories "elegant," as cosmologists like to say, but we need more than that for science. We need empirical tests. Still, the history of science compels me to embrace the multiverse concept for the simple reason that this is the next natural stage in our scientific development. We went from the solar system being all there is, to the galaxy being all there is, to the universe being all there is. Like unsheathing Russian nested dolls, there is no reason to think that there is not another cosmic shell to pull off, and thus I give this heresy a fuzzy factor of .7.

Heresy 2. Time Travel Is Possible
Although time travel is the staple of science fiction writers, most physicists and cosmologists agree that time travel is impossible. Not only does

it violate numerous physical laws, there are fundamental problems of causality. The most prominent is the "matricide paradox" in which you travel back in time and kill your mother before she had you, which means you could not have been born to then travel back in time to kill your mother. In *Back to the Future*, Marty McFly faces a related but opposite dilemma in which he must arrange for his parents to meet in order to ensure his conception. In the original *Star Trek* series Dr. McCoy falls through a time portal and changes the past in a way that erases the *Enterprise* and her crew, with the exception of Kirk and Spock, who must travel back to fix what McCoy has undone.

One way around such paradoxes can be found in extremely sophisticated virtual reality machines (think of *Star Trek*'s holodeck machine), programmed to replicate a past time and place in such detail that it would be nearly indistinguishable from the real past. Another theory involves the parallel universes model of the multiverse in which you travel back in time to a *different* universe from your own, although you would select one very similar so as to experience a past very much like that of our own universe. This was the plot line in Michael Crichton's science fiction novel *Time Line*, in which the characters journeyed back to medieval Europe in a closely parallel universe, without concern for mucking up their own time line.

The fundamental shortcoming for both of these time travel scenarios is that it isn't *really* your past. A virtual reality is simply a museum writ large, and transporting to some other universe's past would be like going back and meeting your mother only to witness her marry someone other than your father, thus eliminating you from that past—surely a less appealing trip than one in your own time line. To do that you need the time machine of the California Institute of Technology cosmologist Kip Thorne, who had his interest in time travel piqued when he received a phone call one day from the Cornell University astronomer Carl Sagan, who was looking for a way to get the heroine of his novel *Contact*—Eleanor Arroway (played by Jodie Foster in the film version)—to the star Vega twenty-six light-years away.

The problem Sagan faced, as all science fiction writers do in such situations, is that at the speed of spacecraft today it would take over 300,000 years to get to Vega. That's a long time to sit, even if you are in first class with your seat back and tray table down. Thorne's solution, adopted by

Sagan, was to send Ellie through a wormhole—a hypothetical space warp similar to a black hole in which you enter the mouth and fall through a short tube in hyperspace that leads to an exit hole somewhere else in the universe (think of a tube running through the middle of a basketball—instead of going all the way around the surface to get to the other side, you take a short cut through the middle). Thorne's calculations showed that it was theoretically possible for Ellie to travel just one kilometer down the wormhole tunnel and emerge near Vega moments later—not even time for a bag of peanuts.

Since, as Einstein showed, space and time are intimately entangled, Thorne theorizes that by falling through a wormhole in one direction it might be possible to travel backward in time, while traversing the wormhole in the other direction one could go forward in time. Thus, a wormhole would be a time warp as well as a space warp, making it a type of time machine. After publishing his theory in a technical physics journal in 1988, the media got hold of the story and Kip was branded "The Man Who Invented Time Travel." Not one to encourage such sensationalism, Thorne continued his research and by the early 1990s began growing skeptical of his own thesis. Whether it is possible to actually travel through a wormhole without being crushed out of existence, Thorne reasoned, depends on the laws of quantum gravity, which are not fully understood at this point. Even his colleague Stephen Hawking had serious doubts. As far as we know, Hawking noted only half sardonically, the universe contains a "chronology protection conjecture," which says that "the laws of physics do not allow time machines," thus keeping "the world safe for historians." In addition, if time travel were possible we should be getting frequent visits from future travelers! Where are all the time tourists, Hawking wondered? It's a good question and, in conjunction with the causality paradoxes and physical law constraints, makes me skeptical as well; until much more is known about quantum gravity and wormholes, virtual reality and the multiverse, I assign this a fuzzy factor of .3.

Heresy 3. Evolution Is Not Progressive

That evolution happened is about as close to factual as a scientific theory can get, but there is an assumption about evolution that is so pervasive that even advertising agencies know they can count on the public's acceptance of it. You can see it in a full-page ad that ran in the *New York Times* on

June 3, 2001. "THE YELLOW PAGES ARE EVOLVING" reads the copy beneath what is unquestionably *the* icon of progressive evolution: humans evolved from apes (specifically the chimpanzee) and in the process got taller and smarter. The image itself was made famous by the endlessly reproduced foldout page from a Time/Life book on evolution, but it represents a deeper bias about how evolution works that is wrong on two levels.

First, humans did not evolve from chimpanzees or any other modern ape. Humans and modern apes evolved from a common ancestor we all shared about six million years ago. Although it is true that modern humans are taller and have bigger brains than primates did millions of years ago, evolution has been anything but progressive and linear in this straight-line fashion. In fact, it now appears that throughout the past six million years dozens of hominid species existed, many of them living simultaneously. We know that Neanderthals shared much of Europe and the Middle East with modern humans, that they were bigger than us, had a reasonably complex tool kit, wore clothes, and had brains slightly larger than our own. If evolution is progressive, why are there no Neanderthals around today? The answer is probably language (we had it, they didn't), but the point is that evolution is best described not as a ladder of progress with each more advanced species on the next rung up, but as a richly branching bush, with each branch and twig shooting off in no particular direction, and no species more "advanced" than any other. The reason is that evolution does not look to the future. Its guiding principle is the here and now—local adaptations to local environments in order to get your genes into the next generation. There is no long-term goal or progressive purpose to evolution.

Second, the bias of progressive evolution can be seen in the grand description of life's history from bacteria to brains. Wipe out all forms of big and complex life today through a catastrophic meteor impact or nuclear war and, the progressive bias dictates, in a couple of billion years the surviving bacteria living in rocks deep below the earth's surface will emerge and reevolve into something like us again. The myth buster here is Stephen Jay Gould, the late Harvard paleontologist whose numerous books and popular essays on evolution have consistently chipped away at the belief in inherent evolutionary progress, proposing in its stead a theory of evolution that says we are nothing more than a glorious accident, a contingent quirk on life's landscape. Rewind the tape of life and play it back, Gould says, and nothing like humans would evolve again (the tape is a metaphor, of course, as a Read Only Memory tape would play back exactly as it was originally recorded). The trillion evolutionary steps between basic bacteria and big brains would not happen again in a million playbacks of the film of life. We are unique, literally a once-in-a-lifetime fluke of nature.

To be fair, I must confess a personal preference for this gloriously contingent view of life, so my fuzzy factor may be too high. It is certainly higher than many evolutionary theorists like Richard Dawkins or Daniel Dennett would give it, since they see in the pattern of evolution a story of upward progress to ever greater heights and smarts where, if the tape were rewound, another intelligent communicating species would arise. Maybe, but in the 3.5-billion-year history of life that included hundreds of millions of species, only one has made it—us—so I give this heresy a fuzzy factor of .8.

Heresy 4. Oil Is Not a Fossil Fuel

Ever since the energy crisis of the 1970s we have been hearing about how the world is about to run out of such fossil fuels as natural gas and oil, and that we must quickly develop alternative sources. The original doomsayers appear to be wrong since, thirty years on, gas-guzzling SUVs are chugging around and getting their tanks filled at rates well below what inflation has done to most other consumer prices. Still, environmentalists argue, whether fossil fuels run out next year or next decade, they are going to run out relatively soon.

Not if Cornell University's Thomas Gold is right in his radical theory about the "deep hot biosphere." Miles below the surface of the earth, deeply embedded in the nooks and crannies of rocks, live primitive thermophilic (heat-loving) bacteria, similar to the microorganisms that have been discovered living in thermovents deep on the ocean bottom and in the hot pools of Yellowstone National Park, and as tough as the extremophiles living within ice sheets in Antarctica and possibly even on meteors in deep space. This deep hot biosphere gets its energy not from the sun like we surface-dwelling creatures do, but from the energy of the earth's interior. It may be so dense that it cumulatively outweighs all surface life combined, including trees and plants!

Fossil fuels, or hydrocarbons, says Gold, are not the by-product of decaying organic matter as most geologists believe. Instead, long before

life formed on earth hydrocarbons developed naturally in the planet's interior, just as they have been discovered on other planetary bodies and
moons in the solar system. From the light gas methane to the heavy liquid
petroleum, hydrocarbons exist in prodigious quantities and great depths
and could sustain our energy needs for many centuries or millennia to
come (George Dubya Bush would love this theory).

The reason scientists think that hydrocarbons have their origin in dead
plants is that petroleum contains molecules that are typically the by-product
of decaying organic matter. Also, when you pass light through petroleum it
exhibits an optical property of rotating in a right-handed fashion, which is
the result of having more right-handed molecules than left-handed molecules. (Molecules come in both right- and left-handed versions, but living
organisms consist mostly of right-handed molecules, so a preponderance
of them would indicate an organic origin.) The reason for this, says Gold,
is that petroleum and other hydrocarbons have seeped up through the
rocks from tens of kilometers below the surface, and in so doing have
absorbed organic matter along the way. These organic signs, he concludes,
are secondary to the true origin of hydrocarbons.

Evidence for Gold's theory comes from numerous sources: petroleum
from deeper levels in the crust contains fewer signs of biological origin
than petroleum from shallower levels; oil from different regions of the
planet should show differing chemical signs because of the different
forms of life from which it was allegedly formed, yet all oil shows a common chemical signature, which you would expect if it had a common origin deep inside the earth; one would expect to find oil at geological levels
of abundant plant life but, in fact, it is found below such layers; the natural gas methane is found in many locations where life most likely did not
thrive; diamonds are carbon crushed under high pressure, which Gold
thinks implies the presence of carbon hundreds of kilometers below the
surface.

Perhaps most striking, Gold notes that most oil fields contain far more
reserves than oil companies anticipated because, he argues, they are
refilled from the much larger hydrocarbon supply lying below—the drop
in pressure in the oil cavity caused by drilling draws the hydrocarbons
from the higher-pressure cavities below. Finally, the earth's surface is very
rich in carbonate rocks, which, as their name implies, are loaded with
carbon. Gold believes that the source of the carbon is not biological but

astronomical—the earth was formed by an accretion of rocks similar to the meteorites that bombard the planet today (so-called shooting stars), one type of which is a carbonaceous chondrite. When heated under the extreme pressure of a condensing earth they would have released substantial quantities of hydrocarbons. Lighter than the surrounding material, they would then rise toward the surface, thus accounting for the high carbon content of the earth's crust.

Geologists and earth scientists have explanations for these anomalies, but I'm impressed with Gold's track record of being right about other heresies he has proposed, such as the nature of pulsar stars, the extent of the layer of moon dust the Apollo astronauts would encounter, and even a new theory of hearing. Like any good scientist, Gold admits that the only way to find out if his theory is correct is to drill deep into the earth's surface and see what's down there. In other words, the theory will stand or fall on the evidence. Although my Caltech colleagues and other geologists I have consulted remain extremely skeptical, Gold's logic and evidence lead me to give this heresy a fuzzy factor of .5.

Heresy 5. Cancer Is an Infectious Disease

Have you ever wondered why it is that if cancer and heart disease are the result of genetics, diet, and age, young, healthy, strong people on good diets occasionally drop dead of cancer and heart disease at an early age? Amherst College medical researcher Paul Ewald thinks he knows why— some cancers, heart disease, and other chronic illnesses are the result of infections. You can "catch" cancer, says Ewald. You can down boxes of granola and jog a hundred miles a week and still drop dead of a heart attack before you're forty—witness Jim Fixx and other exercise gurus and diet fanatics who did not live long enough to witness the seemingly ageless George Burns smoke, drink, and womanize his way to age ninety-nine.

There is irony in this heresy because it falls under the umbrella of the germ theory of disease, not accepted until well into the nineteenth century and resisted for decades by leading medical researchers and practitioners (before the germ theory, for example, medical experts believed that malaria—Italian for "bad air"—was caused by malodorous fumes, not the bite of a protozoa-infested mosquito). Once the germ theory was embraced, however, the search for deadly pathogens raged throughout the twentieth century. So why not apply the theory to cancer and heart disease? If you think about it, this is not such a shocking idea. It was not so long ago that the accepted medical dogma was that peptic ulcers were caused by stress. We now know that a bacterium named *Helicobactor pylori* causes peptic ulcers, and it does its nasty work whether you are on Wall Street or Easy Street. Now, surely not *all* cancers are caused by viruses, but Ewald notes the connection between the hepatitis virus and liver cancer, and the papillomavirus and cervical cancer, and these associations between virus and cancer are no longer considered so heretical. So why not other virus-cancer associations, or even virus-whatever connections? Ewald notes studies connecting *Chlamydia pneumoniae* to both atherosclerosis and Alzheimer's, and the infectious protozoan *Toxoplasma gondii* to schizo-phrenia. Might there be more? There are, says Ewald, a lot more.

Of course, we cannot overlook the fact that more older people get cancer, atherosclerosis, and Alzheimer's, so obviously infectious agents alone cannot be the cause. As with most matters medical, there is an interaction between variables; in this case aging bodies with older and weaker immune systems are more susceptible to the actions of nasty germs. But it is the germs we should focus on in terms of control or eradi-cation, says Ewald, not these secondary conditions.

Ewald's heresy springs from the new science of "evolutionary medi-cine," an approach recognizing that viruses, like people, evolve—witness HIV. The drug cocktails work only temporarily because the virus evolves strategies to get around the treatments. It is a constant battle between the host's immune system and the pathogen's survival strategies. Consider the herpes simplex virus that gives you cold sores. It evolved a clever strategy of living in your nervous system so that your body's immune system cannot attack it (without also destroying the nerves, and in the process kill its host). To perpetuate itself, the herpes virus travels down the length of your nerves to endings in moist areas like your mouth and lips where, when you are under stress (during which your immune system is busy doing other things), it causes an open wound through which it can spread to new hosts through such activities as kissing.

In a similar fashion, the human papillomavirus hides in the genital tract and erupts in a herpes blister in the cervix, awaiting sexual contact. Even worse, the papillomavirus takes over cells of the cervix, causing them to replicate at a much higher rate than normal (in order to more rapidly spread itself by injecting its DNA into these host cells). A by-product of this accelerated cellular replication is that some mutate into cancerous cells that then spread. By the time the cancer kills the host, the papillomaviruses have spread to other hosts through sex, and so on indefinitely into the future. That is how evolution works, whether you are a virus or a human. Cancer is simply collateral damage of the virus's survival strategy.

Like everything associated with the theory of evolution, evolutionary medicine is controversial and remains on the heretical fringes of mainstream science. Nevertheless, the human genome project has shown us that the evolutionary perspective cannot be ignored in any aspect of the human condition, and thus I give this heresy a fuzzy factor of .6.

Heresy 6. The Brain and Spinal Cord Can Regenerate
During the 2001 Super Bowl the actor Christopher Reeve generated considerable controversy when he appeared in a commercial that depicted him walking, despite a debilitating spinal cord injury that turned him into a quadriplegic. This sent a false message of hope, numerous advocates for the handicapped argued, because everyone knows that the brain and spinal cord cannot be regenerated. After a couple of weeks—months at most—injuries to the brain and spinal cord are permanent and irrecoverable.

This has been the medical litany for decades, but it may soon be changing. Neuroscientists around the world are racing to capture what would be one of the greatest prizes in medical history—repairing damaged brains and spinal cords, thereby releasing hundreds of thousands of people from the neuroimprisonment of quadri- and paraplegia, Alzheimer's, Huntington's, Parkinson's, Lou Gehrig's, and other neurological diseases. Ira Black, a neuroscientist at the Robert Wood Johnson Medical School, documents this pathbreaking research in his compelling book *The Dying of Enoch Wallace: Life, Death, and the Changing Brain*. The story began fifty years ago in fascist Italy when Rita Levi-Montalcini discovered a hormone that stimulates nerves to grow in size, length, and number. Since then considerable research on nerve growth factor (NGF) has led to the discovery of numerous chemicals that appear to stimulate nerves to grow. Chan-Nao Liu and William Chambers at the University of Pennsylvania Medical School in Philadelphia, for example, documented new neural growth in the cut spinal cords of cats—not from the damaged nerves but from nerves adjacent to those cut, sprouting like tree branches in search of a match on the other side of the cut cord. Oxford University's

Geoffrey Raisman recorded the regrowth of synaptic connections between neurons in the brain (the synapse is the gap between two neural connections), where the destruction of one synaptic connection triggered nerve fibers from undamaged sites to rush in and reconnect the neurons. After more than a hundred experiments Raisman concluded that "the central nervous system can no longer be considered incapable of reconstruction in the face of damage."

Although most of this research is done with animals, it never hurts to remind ourselves that we, too, are animals subject to the same laws of biochemistry. Thus, it is encouraging to read about Fernando Nottebohm's discovery that songbirds generate thousands of new neurons in their brains every day. Birds are not mammals, but at Rockefeller University Elizabeth Gould discovered that rats whose adrenal glands are removed promptly lose massive numbers of neurons in their hippocampus . . . and just as rapidly replace the lost neurons with new ones. Now at Princeton, Gould demonstrated a similar regeneration effect in adult tree shrews, as well as in marmosets (a New World monkey), which, relative to birds and rats, is closely related to humans. Most important, Gould found this neurogenesis (new nerve growth) effect in the monkeys' neocortex, the brain structure responsible for complex thought, and that the "use it or lose it" principle normally applied to muscle development appears to have applications to nerve generation.

The best hope for neurogenesis in humans, however, is in stem cell research, recently the subject of contentious public debate, with President George W. Bush navigating a delicate course through the ragged shoals of secular and religious pundits arguing for and against what promises to be the medical miracle of the twenty-first century. Stem cells are undifferentiated cells awaiting final instructions on what they will be when they grow up. Under the right conditions, it might be possible to program these juvenile cells to develop into neurons destroyed by Parkinson's, cells lost to diabetes, or whatever ails you. Italian scientist Angelo Vescovi, for example, found that he could grow brain tissue in a jar from a handful of stem cells taken from a mouse. It remains to be seen if they can successfully be put back into the brain, but Salk Institute geneticist Fred Gage has a promising study in which he autopsied five cancer victims who had received a chemical marker called bromodeoxyuridine (BrdU) that is used to track how many new cells are being created in the body (chemotherapy destroys *all* duplicating cells in the body, but since cancer cells replicate faster, the hope is that the chemo kills them all before killing the patient). To his astonishment Gage discovered BrdU in primitive neural stem cells in the brain, indicating that these elder people were generating new neurons, perhaps as many as five hundred to a thousand a day. It would appear that you *can* teach an old dog new tricks.

A subject pool of five is limited, to be sure, and skeptics of animal studies abound—Yale's Pasko Rakic, for example, thinks that neurogenesis is less common among bigger-brained mammals because replacing neurons means replacing memories, which are critical for higher complex thought. Even if true, however, I am optimistic that stem cells are the key to an artificial form of human neurogenesis. It will probably be many more years—decades more likely—before Christopher Reeve can don his Superman cape and leap tall buildings in a single bound, and perhaps there is an element of wishful thinking on my part after nursing a paraplegic girlfriend through years of rehab that still resulted in paralysis, but the remarkable progress made in neuronal growth factors, spinal cord regeneration in higher mammals, and especially in stem cell research leads me to give this heresy a fuzzy factor of .7.

Heretics and Skeptics

From our perspective in the twenty-first century, it is easy to look back to the time of Copernicus or Darwin and think, "How could they have *not* realized that the earth goes around the sun and that life evolved?" And yet the pre-Copernican and pre-Darwinian worldviews were as real for those folks as our scientific worldviews are for us. Can we really afford the smug self-aggrandizement that wells up whenever we lose our historical perspective when, four centuries from now, our descendants may laugh at the risible notions we accept as factual today?

Whether these heresies float or sink upon the turbulent waters of scientific debate is not the deeper message in this exercise in skepticism. Science, if we think of it as a set of methods to answer questions about nature instead of a body of facts to be dogmatically distilled, is intimately dependent upon its heretics and skeptics who have the courage and insight to challenge the status quo. Most heresies in science do not survive—for every Galileo, Darwin, and Einstein who shattered the pillars of knowledge there are a thousand long-forgotten names whose creative speculations tanked. But those that triumph give us a whole new way of looking at the world, and offer a completely different set of questions to answer. All revolutions in science stem from heretics and skeptics, and thus even though most of them are probably wrong most of the time, we must keep an open mind because one never knows where and when the universe will once again explode in revolutionary change.

4

The Virtues of Skepticism

A Way of Thinking and a Way of Life

ON WEDNESDAY, AUGUST 9, 2000, I appeared as a guest on Boston's WTKK, 96.9 FM talk radio, hosted by a genial but verbose woman named Jeanine Graf. The interview was set up by my publicist at the University of California Press to promote my just-released book on Holocaust denial, *Denying History*, but since presidential candidate Al Gore had just selected the Orthodox Jewish senator Joseph Lieberman as his vice presidential running mate, the subject everyone wanted to discuss was politics and religion.

Most callers were impressed and pleased with Gore's choice, many extolling the virtues of biblical ethics and how it is good that our politicians not only endorse their favorite biblical characters (Bush identified Jesus as the philosopher who most influenced his life, while Gore called himself a born-again Christian whose deep faith sustains him), but that they actually reintroduce biblical ethics into politics. To one caller I responded: "Oh, do you mean such biblical ethical practices as stoning to death disobedient children?" The caller promptly challenged me to produce the said passage. As I was nowhere near a Bible, he said that if I could post it to the Skeptics Society Web page (www.skeptic.com) within the next twenty-four hours he would donate $100. If I could not produce the passage, then I had to donate $100 to his favorite charity, a Jewish organization called Jews for the Protection of Firearm Ownership.

The host of the show took the caller's phone number and insisted that we actually play out this little bet, and that she would have the two of us on the show the next night to settle it. The next morning I phoned the religion editor of *Skeptic* magazine, Tim Callahan, whose two books—*Bible Prophecy* and *The Secret Origins of the Bible*—have pushed him to the forefront of biblical scholarship. I figured that Tim would know where that passage came from and, sure enough, a few minutes later he rang back with the answer—Deuteronomy 21:18–21 (Revised Standard Version):

> If a man has a stubborn and rebellious son, who will not obey the voice of his father or the voice of his mother, and, though they chastise him, will not give heed to them, then his father and his mother shall take hold of him and bring him out to the elders of his city at the gate of the place where he lives, and they shall say to the elders of his city, "This our son is stubborn and rebellious, he will not obey our voice; he is a glutton and a drunkard." Then all the men of the city shall stone him to death with stones; so you shall purge the evil from your midst; and all Israel shall hear, and fear.

Within an hour my webmeister, Nick Gerlich, had the passage up on our Web page, and early that evening I found myself back on the air with Graf and my challenger. I fully expected that either he would not appear on the show or he would waffle and not pay up. I was wrong on both counts. Not only was he on the air, he spoiled Graf's hope for a good talk-radio on-air confrontation when he confessed: "There is nothing to dispute. I was wrong. I will pay the Skeptics Society one hundred dollars." And he did. I was blown away. This almost never happens. It takes considerable intellectual courage and honesty to admit you are wrong, and this gentleman did so with grace. He did wish to make a couple of other points, which included the fact that Jesus' new philosophy of ethics overrode much of Old Testament morality by calling for ecumenical acceptance of those outside of our group and more humane responses to sinners and wrongdoers. (I even received a call the next morning from a woman making the same point, who also kindly offered to pray for me to accept Jesus as my savior.)

I agreed with the radio caller, but pointed out that Christians typically pick and choose biblical passages without consistency, including and especially from the Old Testament, and that there are many more Old Testament rules that make one blanch and feel embarrassed for believers. For example, for emancipated women thinking of adorning themselves in business attire that may resemble men's business ware (or for guys who dig cross-dressing), Deuteronomy 22:5 admonishes: "A woman shall not wear anything that pertains to a man, nor shall a man put on a woman's garment; for whoever does these things is an abomination to the Lord your God."

Even worse than stoning disobedient children (for it also encompasses a wide range of misogynistic attitudes) is how to deal with virginal and nonvirginal women. According to Deuteronomy 22:13–21, for all you men who married a nonvirgin, you've got to turn in your wife immediately for a proper stoning. In the following passage, for those not accustomed to reading between the biblical lines, the phrase "goes in to her" should be taken literally, and "the tokens of virginity" means the hymen and the blood on the sheet from a virgin's first sexual experience; the key passage about stoning her to death is at the end.

> If any man takes a wife, and goes in to her, and then spurns her, and charges her with shameful conduct, and brings an evil name upon her, saying, "I took this woman, and when I came near her, I did not find in her the tokens of virginity," then the father of the young woman and her mother shall take and bring out the tokens of her virginity to the elders of the city in the gate; and the father of the young woman shall say to the elders, "I gave my daughter to this man to wife, and he spurns her; and lo, he has made shameful charges against her, saying, 'I did not find in your daughter the tokens of virginity.' And yet these are the tokens of my daughter's virginity." And they shall spread the garment before the elders of the city. Then the elders of that city shall take the man and whip him; and they shall fine him a hundred shekels of silver, and give them to the father of the young woman, because he has brought an evil name upon a virgin of Israel; and she shall be his wife; he may not put her away all his days. But if the thing is true, that the tokens of virginity were not found in the young woman, then they shall bring out the young woman to the door of her father's house, and the men of her city shall stone her to death with stones, because she has

wrought folly in Israel by playing the harlot in her father's house; so you shall purge the evil from the midst of you.

Similarly, for those who have succumbed to the temptation of the flesh at some time in your married life, Deuteronomy 22:22 does not bode well: "If a man is found lying with the wife of another man, both of them shall die, the man who lay with the woman, and the woman; so you shall purge the evil from Israel." Do Jews and Christians *really* want to legislate biblical morality, especially in light of the revelations of the past couple of decades of the rather low moral character of many of our religious leaders? Most don't, but believe it or not some do, even advocating returning to stoning as a proper form of punishment.

(To the religious right who lobby for the Ten Commandments to be posted in public schools and other public venues, please note that the very first one prohibits anyone from believing in any of the other gods besides Yahweh. "Thou shall have no other gods before me" is a passage indicating that polytheism was commonplace at the time, and that Yahweh was, among other things, a jealous god. By posting the Ten Commandments, we are sending the message that any nonbeliever, or believer in any other god, is not welcome in our public schools. This is not an attitude in keeping with the U.S. Constitution and is rightly prohibited by the First Amendment.)

To be fair to believers, not all biblical ethics are this bad. There is much to pick and choose from that is useful for our thinking about moral issues. The problem here is consistency, and selecting ethical guidelines that support our particular personal or social prejudices and preferences. If you are going to claim the Bible as your primary (or only) code of ethics, and proclaim (say) that homosexuality is sinful and wrong because the Bible says so, then to be consistent you've got to kill rebellious youth and nonvirginal premarried women. Since most would not endorse such an ethic, why target gays and lesbians but cut some slack for rebellious youths and promiscuous women? And, on the consistency issue, why aren't men subject to the same set of sexual guidelines as women? The answer is that in that culture at that time it simply was not appropriate. Thankfully we have moved beyond that culture. What we really need is a new set of morals and an ethical system designed for our time and place, not one scripted for a pastoral/agricultural people who lived four thousand years ago. We should think through these moral issues for ourselves instead of turning to what is largely an antiquated book of morals.

This is, in fact, one of the primary goals of the modern skeptical movement that has grown dramatically over the past quarter century (and the subject of my book *The Science of Good and Evil*). Skepticism, of course, dates back to the ancient Greeks, well captured in Socrates' famous quip that all he knows is that he knows nothing. Skepticism as nihilism, however, gets us nowhere and, thankfully, almost no one embraces it. The word *skeptic*, in fact, comes from the Greek *skeptikos*, for "thoughtful"— far from modern misconceptions of the word as meaning "cynical" or "nihilistic." According to the *Oxford English Dictionary*, *skeptical* has also been used to mean "inquiring," "reflective," and, with variations in the ancient Greek, "watchman" or "mark to aim at." What a glorious meaning for what we do! We are thoughtful, inquiring, and reflective, and in a way we are the watchmen who guard against bad ideas, consumer advocates of good thinking who, through the guidelines of science, establish a mark at which to aim.

Since the time of the Greeks, skepticism (in its various incarnations) has evolved along with other epistemologies and their accompanying social activists. The Enlightenment, on one level, was a century-long skeptical movement, for there were no beliefs or institutions that did not come under the critical scrutiny of such thinkers as Voltaire, Diderot, Rousseau, Locke, Jefferson, and many others. Immanuel Kant in Germany and David Hume in Scotland were skeptics' skeptics in an age of skepticism, and their influence continues unabated to this day (at least in academic philosophy and skepticism). Closer to our time, Charles Darwin and Thomas Huxley were skeptics par excellence, not only for the revolution they launched and carried on (respectively) against the dogma of creationism, but also for their stand against the burgeoning spiritualism movement that was sweeping across America, England, and the Continent. Although Darwin was quiet about his spiritual skepticism and worked behind the scenes, Huxley railed publicly against the movement, bemoaning it in one of the great one-liners in the history of skepticism: "Better live a crossing-sweeper than die and be made to talk twaddle by a 'medium' hired at a guinea a séance." In the twentieth century Bertrand Russell and Harry Houdini stand out as representatives of skeptical thinkers and doers (respectively) of the first half, and skepticism in the second half of the century was marked by Martin Gardner's *Fads and Fallacies in the Name of Science*, launching what we think of today as "the skeptics."

No movement or institution leaps into existence out of a sociohistorical vacuum, spontaneously erupting like a universe out of a quantum foam fluctuation, but we have to start a historical sequence somewhere, so I date the modern skeptical movement to 1950 with the publication of an essay by Martin Gardner in the *Antioch Review* titled "The Hermit Scientist." The essay is about what we would today call pseudoscientists, and it was Gardner's first-ever publication of a skeptical nature. It not only launched a lifetime of critical analysis of fringe claims, but in 1952 (at the urging of Gardner's literary agent, John T. Elliott) the article was expanded into a book-length treatment of the subject under the title *In the Name of Science*, with the descriptive subtitle, "An entertaining survey of the high priests and cultists of science, past and present." Published by Putnam, the book sold so poorly that it was quickly remaindered and lay dormant until 1957, when it was republished by Dover and has come down to us as *Fads and Fallacies in the Name of Science*, still in print and arguably the skeptic classic of the past half century. (Gardner realized his book had made it, he explained, when he turned on the radio "at three A.M. one morning, when I was giving a bottle of milk to my newborn son, and being startled to hear a voice say, 'Mr. Gardner is a liar.' It was John Campbell, Jr., editor of *Astounding Science Fiction*, expressing his anger over the book's chapter on dianetics.")

What caught the attention of a youthful Martin Gardner half a century ago was the "hermit scientist," working alone and usually ignored by mainstream scientists: "Such neglect, of course, only strengthens the convictions of the self-declared genius," Gardner concluded in his original 1950 paper. "Thus it is that probably no scientist of importance will present the bewildered public with detailed proofs that the earth did not twice stop whirling in Old Testament times, or that neuroses bear no relation to the experiences of an embryo in the mother's womb" (referring to L. Ron Hubbard's dianetics theory that negative engrams are imprinted in the fetus's brain while in the womb).

Gardner was, however, half wrong in his prognostications: "The current flurry of discussion about Velikovsky and Hubbard will soon subside, and their books will begin to gather dust on library shelves." While Velikovskians are a quaint few surviving in the interstices of fringe culture,

L. Ron Hubbard has been canonized by the Church of Scientology and deified as the founding saint of a world religion.

In the first chapter of *In the Name of Science*, Gardner picks up where he left off, noting that "tens of thousands of mentally ill people throughout the country entered 'dianetic reveries' in which they moved back along their 'time track' and tried to recall unpleasant experiences they had when they were embryos." Half a century later Scientology has converted those reveries into a worldwide cult of personality surrounding L. Ron Hubbard, targeting celebrities for membership and generating hundreds of millions of dollars in tax-free revenue as an IRS-approved "religion."

Today UFOs are big business, but in 1950 Gardner could not have known that the nascent flying saucer craze would turn into an alien industry: "Since flying saucers were first reported in 1947, countless individuals have been convinced that the earth is under observation by visitors from another planet." Absence of evidence then was no more a barrier to belief than it is today, and believers proffered the same conspiratorial explanations for the dearth of proof: "I have heard many readers of the saucer books upbraid the government in no uncertain terms for its stubborn refusal to release the 'truth' about the elusive platters. The administration's 'hush-hush policy' is angrily cited as proof that our military and political leaders have lost all faith in the wisdom of the American people."

From his perspective in 1950 Gardner was even then bemoaning the fact that some beliefs never seem to go out of vogue, as he recalled H. L. Mencken's quip from the 1920s that "if you heave an egg out of a Pullman car window anywhere in the United States you are likely to hit a fundamentalist." Gardner cautions that when presumably religious superstition should be on the wane how easy it is "to forget that thousands of high school teachers of biology, in many of our southern states, are still afraid to teach the theory of evolution for fear of losing their jobs." Today, Kansas, Texas, Georgia, Ohio, and other states enjoin the fight as the creationist virus spreads nationwide.

I devote an entire chapter in my book *The Borderlands of Science* to Martin Gardner and his seminal work, but suffice it to say here that *Fads and Fallacies in the Name of Science* has been a cherished classic read by legions of skeptics and scientists and laid the foundation for a bona fide skeptical movement that found its roots in the early 1970s. There has been some debate (and much quibbling) about who gets what amount of credit for the founding of the modern skeptical movement in the journal of the Committee for the Scientific Investigation of Claims of the Paranormal (CSICOP), *Skeptical Inquirer* (much of this history has been outlined in the pages of my own magazine, *Skeptic*, in interviews with the leading lights of the skeptical movement). This is not the place to present a definitive history of the movement, but from what I have gleaned from first- and secondhand sources Martin Gardner, magician James Randi, psychologist Ray Hyman, and philosopher Paul Kurtz played primary roles in the foundation and planning of the organization, with numerous others in secondary supporting roles, such as journalist Phil Klass and sociologist Marcello Truzzi.

Regardless of who might be considered the "father" of the modern skeptical movement, everyone I have spoken to (including the other founders) agrees that it was Paul Kurtz more than anyone else who actually made it happen. All successful social movements have someone who has the organizational skills and social intelligence to get things done. Paul Kurtz is that man. But he had a lot of help. First among equals in this capacity is Barry Karr, who impressed me with his organizational genius and plain old hard work. For a social movement to survive it must be able to make the transition from the first generation to the second, and I have no doubt that CSICOP will flourish in the next quarter century thanks to the next generation of skeptics such as Karr and others.

The founding of the Skeptics Society by myself, Pat Linse, and Kim Ziel Shermer in 1992, then, was also not without precedent and historical roots, and while this history has yet to be written, suffice it to say that without the likes of Gardner, Randi, Hyman, and Kurtz there would be no Skeptics Society and *Skeptic* magazine. And what an experience it has been. When they started this movement I was twenty years old and in my third year of college at Pepperdine University, a Church of Christ–based institution located in Malibu and overlooking the Pacific Ocean. Although the site was certainly a motivating factor in my choice of a college, the primary reason I went there was that I was a born-again Christian who took his mission for Christ seriously. I thought I should attend a school where I could receive some serious theological training, and I did. I took courses in the Old and New Testaments, Jesus the Christ, and the writings of C. S. Lewis. I attended chapel twice a week (although truth be told it was required for all students). Dancing was not allowed on campus (the sexual suggestiveness might trigger already-inflamed hormone production to go into overdrive) and we were not allowed into the dorm rooms of members of the opposite sex.

Despite the restrictions it was a good experience because I was a serious believer and thought that this was the way we should behave anyway. But somewhere along the way I found science, and that changed everything. I was considering theology as a profession, but when I discovered that a Ph.D. required proficiency in several languages (Hebrew, Greek, Aramaic, and Latin), knowing that foreign languages was not my strong suit, I switched to psychology and mastered one of the languages of science: statistics. Here (and in research methodology courses) I discovered that there are ways to get at solutions to problems for which we can establish parameters to determine whether a hypothesis is probably right (e.g., rejecting the null hypothesis at the .01 level of significance) or probably wrong (e.g., not statistically significant). Instead of the rhetoric and disputation of theology, there were the logic and probabilities of science. What a difference this difference in thinking makes!

By the end of my first year of a graduate program in experimental psychology at the California State University, Fullerton, I had abandoned Christianity and stripped off my silver ichthus, replacing what was for me the stultifying dogmas of a two-thousand-year-old religion with the worldview of an always changing, always fresh science. The passionate nature of this perspective was enthused most emphatically by my evolutionary biology professor, Bayard Brattstrom, particularly in his after-class discussions at a local bar that went into the wee hours of the morning. Science is where the action was for me.

About this time (the mid-1970s) Uri Geller entered my radar screen. I recall that *Psychology Today* and other popular magazines published glowing stories about him, and reports were afloat that experimental psychologists had tested the Israeli psychic and determined that he was genuine. My adviser—a strictly reductionistic Skinnerian behavioral psychologist named Doug Navarick—didn't believe a word of it, but I figured there might be something to it, especially in light of all the other interesting research being conducted on altered states of consciousness, hypnosis, dreams, sensory deprivation, dolphin communication, and the like. I took a course in anthropology from Marlene Dobkin de Rios, whose research was on shamans of South America and their use of mind-altering plants. It all seemed entirely plausible to me and, being personally interested in the subject (the Ouija board consistently blew my mind), I figured that this was rapidly becoming a legitimate subfield of psychological research. After all, Thelma Moss had a research laboratory devoted to studying the paranormal, and it was at UCLA no less, one of the most highly regarded psychology programs in the country.

Enter James "the Amazing" Randi. I do not recall exactly when or where I first encountered him. I believe it was on *The Tonight Show* when he was demonstrating how to levitate tables, bend spoons, and perform psychic surgeries. Randi didn't convince me to become a full-fledged skeptic overnight, but it got me thinking that if some of these psychics were fakes, perhaps they all were (and if not fakes, at least self-deceived). Herein lies an important lesson. There is little to no chance that we can convince True Believers of the errors of their thinking. Our purpose is to reach that vast middle ground between hard-core skeptics and dogmatic believers—people like me who thought that there might be something to these claims but simply had never heard a good counterexplanation. There are many reasons why people believe weird things, but certainly one of the most pervasive is that most people have never heard a good explanation for the weird things they hear and read about. Short of a good explanation, they accept the bad explanation that is typically proffered. This alone justifies all the hard work performed by skeptics toward the cause of science and critical thinking. It does make a difference.

Fast-forward ten years. My first contact with organized skepticism came in the mid-1980s through the famed aeronautics engineer and human-powered flight inventor Paul MacCready. I originally met Paul through the International Human Powered Vehicle Association, as he was interested in designing them and I was interested in racing them (I had a ten-year career as an ultra-marathon cyclist). One day he phoned to invite me to a California Institute of Technology lecture being hosted by a group called the Southern California Skeptics (SCS). SCS was an offshoot of CSICOP and one of many that had spontaneously self-organized around the country throughout the 1980s. The lectures were fascinating, and because of my affiliation with Paul I got to meet some of the insiders in what was rapidly becoming the "skeptical movement." Paul was acquainted with such science luminaries as physicists Richard Feynman and Murray Gell-Mann and the evolutionary biologist Stephen Jay Gould, as well as with Randi and the magicians Penn and Teller, so it seemed like skepticism was a real happening. In 1987, CSICOP hosted a convention at the Pasadena Civic Center that featured Carl Sagan as their keynote speaker. Carl was so inspiring that I decided to return to graduate school to complete my doctorate.

By the end of the 1980s, however, the Southern California Skeptics folded and the skeptical movement came to a grinding halt in the very place that so desperately needed it. In 1991, I completed my Ph.D., was teaching part-time at Occidental College, and was nosing around for something different to do. I had just published a paper in a science history journal on the Louisiana creationism trial that featured the activities of SCS, who had organized an amicus curiae brief signed by seventy-two Nobel laureates (encouraged by fellow Nobelist Murray Gell-Mann) and submitted to the United States Supreme Court. One of SCS's former volunteer staff members, Pat Linse, heard about the paper, tracked me down, and dropped by to pick up a reprint copy. During that visit she expressed her frustration—and that of many others—that skepticism in Southern California had gone the way of the Neanderthals. Subsequent meetings with her and others inspired us to jump-start the skeptics movement again by launching a new group and bringing in James Randi for our inaugural lecture in March 1992. It was a smashing success as over four hundred people crammed into a three-hundred-seat hall to hear the amazing one astonish us all with his wit, wisdom, and magic.

With that successful event we were off and running. I starting planning a newsletter, but when Pat saw a sample copy of a bicycle magazine I was publishing (*Ultra Cycling* magazine, the publication of the Ultra-Marathon Cycling Association and Race Across America that I had cofounded in the early 1980s), which was sixty-four pages long, perfect bound, with a duotone coated cover, she said that if we could splurge for a skeptical publication of that quality she would generate the appropriate artwork and typography. Since Pat is a professional artist who was working for movie studios generating film posters, she was more than capable of backing up her offer, which I accepted. Our original cover was to feature Randi, and Pat produced a striking portrait of him, but just before publication Isaac Asimov died, so Pat generated a new cover portrait and that became the cover of volume one, number one of what we came to call *Skeptic* magazine. (My originally planned title—*The Journal of Rational*

Skepticism—was voted down by Pat and my wife, Kim Ziel Shermer, who reasoned that shorter is better. They were right.)

If Martin Gardner was the pen of this revolution, then James Randi was its sword, as he was in the trenches fighting the good fight and inspiring us all to maintain the courage of our convictions in the face of overwhelming odds. But now that I am running a sizable organization myself I have come to respect more than ever before what Paul Kurtz has done for our movement. He may not be as prolific and famous a writer as Martin Gardner, or as public and visible an activist as James Randi, but in terms of the day-to-day grind of keeping a movement afloat through the constant battering and assaults that come from variegated sources, there are few who can be compared with Paul Kurtz. So I close with several excerpts from what I still consider to be his finest work, *The Transcendental Temptation*.

The temptation, says Kurtz, "lurks deep within the human breast. It is ever-present, tempting humans by the lure of transcendental realities, subverting the power of their critical intelligence, enabling them to accept unproven and unfounded myth systems." Specifically, Kurtz argues that myths, religions, and claims of the paranormal are lures tempting us beyond rational, critical, and scientific thinking, for the very reason that they touch something in us that is sacred and important—life and immortality: "This impulse is so strong that it has inspired the great religions and paranormal movements of the past and the present and goaded otherwise sensible men and women to swallow patently false myths and to repeat them constantly as articles of faith." What drives this temptation? The answer Kurtz provides is both insightful and elegant.

> Let us reflect on the human situation: all of our plans will fail in the long run, if not in the short. The homes we have built and lovingly furnished, the loves we have enjoyed, the careers we have dedicated ourselves to will all disappear in time. The monuments we have erected to memorialize our aspirations and achievements, if we are fortunate, may last a few hundred years, perhaps a millennium or two or three—like the stark and splendid ruins of Rome and Greece, Egypt and Judea, which have been recovered and treasured by later civilizations. But all the works of human beings disappear and are forgotten in short order. In the immediate future the beautiful clothing that we adorn ourselves with, eventually even

our cherished children and grandchildren, and all of our possessions will be dissipated. Many of our poems and books, our paint-
ings and statues will be forgotten, buried on some library shelf or in
a museum, read or seen by some future scholars curious about the
past, and eventually eaten by worms and molds, or perhaps consumed by fire. Even the things that we prize the most, human intelligence and love, democratic values, the quest for truth, will in time
be replaced by unknown values and institutions—if the human
species survives, and even that is uncertain.

Although Kurtz sounds like a pessimist, he's actually a realist, occasionally even an optimist.

Were I to take an inventory of the sum of goods in human life, they
would far outweigh the banalities of evil. The pessimist points to
Caligula, Attila, Cesare Borgia, Beria, or Himmler with horror and
disgust; but I would counter with Aristotle, Pericles, da Vinci, Einstein, Beethoven, Mark Twain, Margaret Sanger, and Madame
Curie. The pessimist points to duplicity and cruelty in the world; I
am impressed by the sympathy, honesty, and kindness that are manifested. The pessimist reminds us of ignorance and stupidity; I, of
the continued growth of human knowledge and understanding. The
pessimist emphasizes the failures and defeats; I, the successes and
victories in all their glory.

The most important point Kurtz makes in *The Transcendental Temptation* comes toward the end in his discussion of the meaning and goals of
skepticism. It is an admonition we should all bear in mind, a passage to be
read once a year.

[I]f there are any lessons to be learned from history, it is that we
should be skeptical of all points of view, including those of the skeptics. No one is infallible, and no one can claim a monopoly on truth
or virtue. It would be contradictory for skepticism to seek to trans-
late itself into a new faith. One must view with caution the promises of any new secular priest who might emerge promising a brave
new world—if only his path to clarity and truth is followed. Perhaps
the best we can hope for is to temper the intemperate and to tame
the perverse temptation that lurks within.

Amen, brother!

SCIENCE AND THE MEANING OF BODY, MIND, AND SPIRIT

5

Spin-Doctoring Science

Science as a Candle in the Darkness

of the Anthropology Wars

THERE IS A MAXIM anthropologists often cite about the geopolitics of diplomacy and warfare among indigenous peoples: the enemy of my enemy is my friend. In reality, of course, the maxim applies to virtually all groups, from tribes and villages to city-states and nation-states—recall the temporary friendship between the United States and the Soviet Union from 1941 to 1945 that promptly dissolved into the cold war upon the defeat of the common enemy.

I thought of this maxim on Monday, November 20, 2000, when I interviewed journalist Patrick Tierney for the science edition of NPR affiliate KPCC's *Airtalk*, when he was in Los Angeles on tour for his just published book *Darkness in El Dorado: How Scientists and Journalists Devastated the Amazon.*[1] Tierney had just flown in from San Francisco where, the previous day, he was pummeled by a panel of experts in front of a thousand scientists gathered at the annual meeting of the American Anthropological Association.

Among the many scientists Tierney goes after, none takes more hits than the anthropologist Napoleon Chagnon, whose study of the Yanomamö people of Amazonia is arguably the most famous ethnography since Margaret Mead's Samoan classics. Since I knew Chagnon's reputation as an intellectual pugilist who has accumulated a score of enemies over the decades, I fully expected that, in obedience to the maxim, they would have rallied around Tierney in a provisional alliance. With a couple of minor exceptions, however, there was almost universal condemnation of the book. A British science writer told *Skeptic* senior editor Frank Miele, who was in attendance: "If I had taken such a beating as Tierney I would have crawled out of the room and cut my throat."[2]

Tierney did seem shell-shocked, as he timorously tiptoed into the studio with a subdued countenance. On the air he seemed almost apologetic for his book, emphasizing that he was not presenting the final word on the subject of the mistreatment of the Yanomamö but, rather, he was merely suggesting the need to investigate the scandalous charges against Chagnon and others that he had gathered in his decade of research.

A wispy thin man with an edge of wilderness about him left over from a waiflike nomadic lifestyle spent chasing down what he thought might be (and his publisher trumpeted as) the anthropological scandal of all time, Tierney struck me as a conciliatory man ill suited for the fight he had instigated. Indeed, he seemed the very embodiment of the type of man Rush Limbaugh would call a bleeding-heart, tree-hugging liberal, and someone environmentalists would call a friend. His publicist at W. W. Norton told me that he was flat broke from years of grant-less, salary-less research and that his very survival hinged on the success of this book. Her interpretation of the AAA meeting, which she attended in hopes of this being a coming-out celebration for a potential bestseller, was that Tierney had few friends there because he was an outsider, a mere journalist at play in the field of science.[3] Who was he to stick his nose in the private business of professionals whose union card—the Ph.D.—was hard earned through the centuries-long system of mentorship and hoop-jumping, not unlike the rites of passage ceremonies young men are put through in many indigenous cultures? Had Tierney simply not paid his scientific dues, or was there something else going on that turned Chagnon's enemies into his friends?

Despite his lack of scientific training Tierney did spend eleven years researching his book, and since outsiders occasionally do make important contributions to science, I wanted to give him a chance. His on-air stories were eye-popping, and his book is filled with so many stories and anecdotes, charges and accusations, backed by interviews, documents, and seventy pages of endnotes and bibliography, that at first blush one is left thinking that if only half or even a tenth of them are true, there is darkness in anthropology, indeed, in all of science.

Darwin's Dictum and Damaged Data

Humans are storytelling animals.[4] Thus, following what I call Darwin's Dictum—"all observations must be for or against some view if they are to be of any service"[5]—we begin by recognizing that Tierney is telling a story against a view he believes has been put forth by certain anthropologists about the Yanomamö and, by implication, about all humanity. Chagnon, he points out, subtitled his best-selling ethnographic monograph on the Yanomamö *The Fierce People*. The French anthropologist Jacques Lizot, Tierney notes, calls the Yanomamö "the erotic people."[6] Chagnon and Lizot, of course, are not immune to the human tendency to dichotomize and pigeonhole, but in telling a story—especially one for or against some view—one is obligated to be fair in properly contextualizing observations and conclusions because the data never just speak for themselves. Thus, the substrate of this essay is the relationship between data and theory, and how journalists and scientists differ in their treatment of that relationship.

For example, Tierney spares no ink in presenting a picture of Chagnon as a fierce anthropologist who sees in the Yanomamö nothing more than a reflection of himself. Chagnon's sociobiological theories of the most violent and aggressive males winning the most copulations and thus passing on their genes for "fierceness," says Tierney, is a Rorschachian window into Chagnon's own libidinous impulses. Chagnon is the bête noire of *Darkness in El Dorado*. In Tierney's pantheon of antiheroes, Chagnon is the anti-Christ of the Yanomamö. The gold miners who kill Yanomamö and destroy their land, and the missionaries who want to "civilize" the Yanomamö by replacing their animistic superstition with a monotheistic one, by comparison, are let off easy.

Indeed, Chagnon is well known in anthropological circles for being tough-minded and occasionally abrasive, and Tierney seemed to encounter no shortage of stories of braggadocio and bellicosity peppered throughout descriptions of a man who himself might best be described as "fierce," in both the jungles of Amazonia and the halls of academia. In a letter to the

Santa Barbara News Press, for example, Chagnon called his critics "so much skunk in the elephant soup":

> I used a metaphor to try to put the nature of academic things into perspective: a soup comprised of one elephant and one skunk. The vast majority of my professional colleagues regard my work with esteem—the elephant part of the soup. . . . But, a highly vocal minority persists in denigrating me and my research in non-academic ways for a variety of reasons, most notably professional jealousy. These represent the acrid flavor that a skunk, even in a very large elephant soup, imparts to it.[7]

In a private e-mail that was forwarded to and published in *Newsweek* (without Chagnon's permission), the embattled anthropologist expressed himself like a true alpha male: "I am encouraged to believe that *The New Yorker* and W. W. Norton [Tierney's excerpter and publisher] are sticking their peckers into a very powerful pickle slicer."[8]

In a two-page account designed to arouse emotion in the reader (it does), Tierney recounts a story told to him by the anthropologist Kenneth Good, who spent twelve years among the Yanomamö (first as a graduate student of Chagnon, then with the German ethologist Iranäus Eibl-Eibesfeldt, and finally with the cultural anthropologist Marvin Harris). Good recalled in a 1995 interview with Tierney that he and Chagnon "used to go down to bars and drink together. It was an embarrassment, but I did it because he was going to be my chair. He was the type of guy who had German shepherd attack dogs, and he'd have people come over to his house in the afternoon and he'd have the students dress up in padded suits and have the dogs attack them. Oh, yes. They'd have to put out an arm or a leg and the dog would attack. Students could get injured."[9] Tierney then turns to Good's book, *Into the Heart*, to retell the story of a violent outburst by Chagnon. Here is how Tierney describes it:

> During his first, nervous night in the jungle, Good was terrified when two screaming men burst inside, pushed him into a table, and ripped his mosquito netting. In the ensuing tussle, all three men wound up sprawling on the ground, bruised and covered with mud, but not before Good recognized his assailants as Chagnon and another anthropologist, both drunk. Good, a tall, husky man, was so angry he threw Chagnon, who is much smaller, over an embankment.

"Tranquilo, Ken," Lizot said, as he helped bring peace.

Fortunately, Chagnon could not remember what had happened to him when he woke up, rather bruised and muddy, the next day. Good never forgot the experience, however. It was the only time anyone ever attacked him in Yanomamiland. "In my twelve years, I witnessed only one raid."[10]

Attack dogs and drunken brawls—it would appear from this narrative that Chagnon is the fierce one, not the Yanomamö. Perhaps, as Tierney argues, even the occasional acts of violence committed by the Yanomamö were nothing more than Chagnon-stimulated outbursts, like something out of *The Gods Must Be Crazy*, where the mere introduction of a Coke bottle disrupts the entire !Kung culture (Chagnon's critics began making this analogy soon after the film's release).[11] As the jacket flap copy for *Darkness in El Dorado* dramatically concludes: "Tierney explores the hypocrisy, distortions, and humanitarian crimes committed in the name of research, and reveals how the Yanomami's internecine warfare was, in fact, triggered by the repeated visits of outsiders who went looking for a 'fierce' people whose existence lay primarily in the imagination of the West."

Tierney's tale above was so inflammatory that I read it aloud to my associates at *Skeptic*, exclaiming about Chagnon, "Can you believe this guy?" I privately wondered whether we all had been duped by him. In fact, *Darkness in El Dorado* is filled with such stories, told mostly in Tierney's words with snippets of partial quotes from his various sources. This literary style always makes me uneasy, so when I interviewed Good I asked him about this incident. He indicated that it happened pretty much as Tierney summarized it, adding that it was more than a little irritating that his mosquito net was torn (malaria-carrying mosquitoes infest the Amazon) and that he was not at all amused by his mentor's inappropriate behavior.[12]

It was with much interest, then, that when Good kindly sent me a copy of *Into the Heart* that I read the original account. Here is Good's rather different description of the event Tierney portrayed as an act of inebriated violence:

> Chagnon, Lizot, and the French anthropologist all knew the Yanomami, and of course their reputation for violence, and having had more than a little to drink, they figured it would be a lot of fun to scare the pants off us. It was, after all, the first night Ray, Eric,

and I were spending in a Yanomami village, and who knew what kinds of fears might be racing through our heads. So they decided they would initiate us. . . .

As Eric and I were busy working with our hammocks and nets, all of a sudden out of the night two big figures burst into the hut screaming, "Aaaaaaaaaahhhhhhh!" grabbing us, and shoving us toward our hammocks, ripping the mosquito netting. My heart skipped a beat. I heard Eric gasp. Bracing myself against a table to keep from falling, I twisted around and saw in the glow of the Coleman Chagnon and the French anthropologist, both of them completely drunk. . . .

Still screaming, I grabbed Chagnon with one arm and the Frenchman with the other and went stampeding out the door with them. There something tripped me up, and I sprawled on the ground, watching as Chagnon and his friend rolled into the eight-foot-deep pit from which the Indians had excavated clay for the hut. Lying there panting, I looked up and saw Lizot emerge from the darkness. "Tranquilo, Ken, tranquilo," he said. "Take it easy, they were just joking."[13]

Note that Tierney leaves off Lizot's qualifier "they were just joking." It was a prank! Tierney turned horseplay into horror. Sure, Good was not amused by the caper, and no doubt alcohol enhanced the pranksters' enthusiasm for playing a practical joke before the long grind of fieldwork was to begin. But regardless of how it is received, a prank is not an "attack" or a "raid."

In my interview with Chagnon he initially called Tierney a "disgusting, slippery, conniving guy," but later reflected that perhaps Tierney's book was simply a case of self-deception, where the author's political agenda of protecting the Yanomamö forced him to misread the data and ethnographies of those he perceived as harming his self-assigned charges.[14] At first I went along with Chagnon in his assessment, as I have witnessed firsthand how powerful self-deception can be among such ideologues as creationists and Holocaust deniers. The more you believe in your own cause, the easier it is to get others to go along. While there may be some self-deception at work here, I fail to see how it can account for the butchering Tierney made of this humorously intended escapade.

Finally, what of Chagnon's "attack dogs"? It turns out that Chagnon is a serious dog trainer—he even wrote a book on the subject titled *Toward the*

Ph.D. for Dogs: Obedience Training from Novice Through Utility, published by Harcourt in 1974. Chagnon was merely demonstrating to his students his highly trained dogs.

The Anthropology Wars

Tierney's book is only the latest in a long line of skirmishes and battles that have erupted in the century-long anthropology wars. The reason such controversies draw so much public attention is that what's at stake is nothing less than the true nature of human nature, and how that nature can most profitably be studied—through rigorous quantitative science or through some other set of methods.

Derek Freeman's lifelong battle with the legacy of Margaret Mead, for example, was not really about whether Samoan girls are promiscuous or prudish. Mead's philosophy (which she inherited from her mentor Franz Boas) that human nature is primarily shaped by the environment was apparently supported by her "discovery" that Samoan girls are promiscuous (because in other cultures promiscuity is taboo and therefore sexual behavior—and by implication all behavior—is culturally malleable). Freeman says Mead was duped by a couple of Samoan hoaxers and had she been more rigorous and quantitative in her research she would have discovered this fact before going to press with what became the all-time anthropological bestseller—*Coming of Age in Samoa*. But, says Freeman, Mead's ideology trumped her science and anthropology lost.[15]

So heated can these debates become that in at least one instance it has led to the fissioning of an academic department. Stanford University now houses the Department of Anthropological Sciences and the Department of Cultural Anthropology. The chairman of the former, Bill Durham, explained to me that the split was not simply between physical and cultural anthropologists, nor is it between those who prefer biological theories of human nature and those who favor cultural theories. "The split is really between those who use and stand behind scientific methods in field and lab work, and those who think science is just another way of knowing, just another paradigm among others. It just so happens that anthropologists often divide on this issue between physical and cultural anthropologists, but not always. There are plenty of cultural anthropologists who conduct rigorous quantitative research. But many others are steeped in postmodernism."[16] Interestingly, Chagnon's ethnography, *Yanomamö*—the epicenter of this whole affair—was published as part of an academic series on "Case Studies in Cultural Anthropology" whose series editors are at Stanford University!

Another venomous snake in the viper pit of the anthropology wars is the question of research ethics. It is simply impossible for anthropologists to observe anything remotely resembling *Star Trek's* "prime directive," where one never interferes with the subject of one's study. To get to know the people, you have to interface with them on numerous levels and no one has ever gotten around the problem of the "observer effect" and retained anything worth saying. That's a given, and the *Code of Ethics* published by the American Anthropological Association is correspondingly vague, offering such "ethical obligations" as:

> To avoid harm or wrong, understanding that the development of knowledge can lead to change, which may be positive or negative for the people or animals worked with or studied.
> To respect the well-being of humans and nonhuman primates.
> To work for the long-term conservation of the archaeological, fossil, and historical records.
> To consult actively with the affected individuals or group(s), with the goal of establishing a working relationship that can be beneficial to all parties involved.

Can you have sexual relations with the natives? *The Code of Ethics* is no help. Point 5 under section A states: "Anthropological researchers who have developed close and enduring relationships (i.e., covenantal relationships) with either individual persons providing information or with hosts must adhere to the obligations of openness and informed consent, while carefully and respectfully negotiating the limits of the relationship."[17] That's as clear as Amazonian mud during the rainy season. Thus, it is hard to say whether the scientists Tierney says were unethical were, in fact, in violation of their professional standards and obligations.

Tierney's strongest case may be against Jacques Lizot who, he documents, engaged in homosexual activities for years with so many Yanomamö young men, and so frequently, that he became known in Yanomamöspeak as "Bosinawarewa," which translates politely as "Ass Handler" and not so politely as "anus devourer."[18] In response to these claims not only did Lizot not deny the basic charges (that also included exchanging goods for sex), but he admitted to *Time* magazine: "I am a homosexual, but my house is not a brothel. I gave gifts because it is part of the Yanomamö culture. I was single. Is it forbidden to have sexual relations with consenting adults?"[19] No, but Tierney disputes both the age of Lizot's partners and whether or not they consented, and suggests that even if it were both legal and moral this is hardly the standard of objectivity one might have hoped for in scientific research, and that it is Lizot who best deserves the descriptive adjective "erotic."

I asked Ken Good about the charges against Lizot. Good said he never once witnessed homosexual behavior in any Yanomamö village and that, in his opinion, it was obvious that the Yanomamö young men were involved with Lizot for one reason only—to obtain machetes and trade goods. I have been unable to resolve any more on the Lizot affair and, in any case, he seems to be a secondary player in this anthropological drama. Despite Tierney's characterization of the Chagnon–Lizot relationship as hostile, Chagnon had no comment at all on Lizot's sexual behavior, and instead told me that "Lizot is a quite capable and thorough scientist, but he's not a particularly good synthetic thinker. He does not always see the bigger picture in his research."[20]

Chagnon, by contrast, is a synthetic, big-picture thinker, and thus it is that the ethics of his research have come under closer scrutiny. Anthropologist Kim Hill from the University of New Mexico, for example, was strongly critical of *Darkness in El Dorado*, yet he expressed his concern about many of the ethical issues the book raises.

> I was concerned about the negative attitude that many Yanomamö I have met seem to have towards Chagnon, and despite the fact that much of this attitude is clearly due to coaching by Chagnon enemies I do believe that some Yanomamö have sincere and legitimate grievances against Chagnon that should be addressed by him. The strongest complaints that I heard were about his lack of material support for the tribe despite having made an entire career (and a good deal of money) from working with them, and his lack of sensitivity concerning some cultural issues and the use of film portrayals. However, I think most of Chagnon's shortcomings amount to little

more than bad judgment and an occasional unwise penchant for self promotion (something which seems to infuriate Yanomamö specialists who are less well known than Chagnon).[21]

Since evolutionary psychologist Steven Pinker published a letter in defense of Chagnon in the *New York Times Book Review* (in response to John Horgan's surprisingly uncritical review of Tierney's book there), I queried him about some of the specific charges. "The idea that Chagnon caused the Yanomamö to fight is preposterous and contradicted by every account of the Yanomamö and other nonstate societies. Tierney is a zealot and a character assassin, and all his serious claims crumble upon scrutiny." What about the charge of ethical breaches? "There are, of course, serious issues about ethics in ethnography, and I don't doubt that some of Chagnon's practices, especially in the 1960s, were questionable (as were the practices in most fields, such as my own—for example, the Milgram studies). But the idea that the problems of Native Americans are caused by anthropologists is crazy. In the issues that matter to us—skepticism, scientific objectivity, classic liberalism, etc.—Chagnon is on the right side."[22]

The carping over minutiae in Chagnon's research methods and ethics that has dogged him throughout his career, however, is secondary to the deeper, underlying issue in the anthropology wars. What Chagnon is really being accused of is biological determinism. To postmodernists and cultural determinists, in calling the Yanomamö "fierce" and explaining their fierceness through a Darwinian model of competition and sexual selection, Chagnon is indicting all of humanity as innately evil and condemning us to a future of ineradicable violence, rape, and war. Are we really this bad? Are the Yanomamö?

Erotic or Fierce?

Anthropology is a sublime science because it deals with such profoundly deep questions as the nature of human nature. This whole "fierce people" business is really tapping into the question of the nature of human good and evil. But to even ask such questions as "Are we by nature good or evil?" misses the complexity of human affairs and falsely simplifies the science behind the study of human diversity. (The propensity to do so is very probably grounded in the tendency of humans to dichotomize the world into unambiguous binary categories.)

Thus, the failure of Tierney's book has less to do with getting the story straight and more to do with a fundamental misunderstanding of the plasticity and diversity of human behavior and a lack of understanding of how science properly proceeds in its attempt to catalog such variation and to generalize from behavioral particulars to categorical universals. Upon finishing the book I let it sit for a couple of days and then plowed through the 147-page rebuttal published on the Internet by the University of Michigan, as well as the many other responses by Chagnon and his colleagues at the University of California, Santa Barbara.[23] I also reread Chagnon's classic work *Yanomamö*. Tellingly, the fourth edition dropped the subtitle *The Fierce People* (although Tierney, characteristically, refers to the book only by the old subtitle). Had Chagnon determined that the Yanomamö were not "the fierce people" after all? No. He realized that too many people were unable to move past the moniker to grasp the complex and subtle variations contained in all human populations, and he became concerned that they "might get the impression that being 'fierce' is incompatible with having other sentiments or personal characteristics like compassion, fairness, valor, etc."[24]

In fact, the Yanomamö call themselves "waiteri" ("fierce") and Chagnon's attribution of them as such was merely attempting "to represent valor, honor, and independence" that the Yanomamö saw in themselves. As he notes in his opening chapter, the Yanomamö "are simultaneously peacemakers and valiant warriors." Like all people, the Yanomamö have a deep repertoire of responses for varying social interactions and differing contexts, even those that are potentially violent: "They have a series of graded forms of violence that ranges from chest-pounding and club-fighting duels to out-and-out shooting to kill. This gives them a good deal of flexibility in settling disputes without immediate resort to lethal violence."[25]

Chagnon has often been accused of using the Yanomamö to support a sociobiological model of an aggressive human nature. Even here, returning to the primary sources in question shows that Chagnon's deductions from the data are not so crude, as when he notes that the Yanomamö's northern neighbors, the Ye'Kwana Indians—in contrast to the Yanomamö's initial reaction to him—"were very pleasant and charming, all of them anxious to help me and honorbound to show any visitor the numerous courtesies of their system of etiquette," and therefore that it "remains true that there are enormous differences between whole peoples."[26] Even on the final page of his chapter on Yanomamö warfare, Chagnon inquires about "the likelihood that people, throughout history, have based their political relationships with other groups on predatory versus religious or altruistic strategies and the cost-benefit dimensions of what the response should be if they do one or the other." He concludes: "We have the evolved capacity to adopt either strategy."[27] These are hardly the words of a hidebound ideologue. In fact, in 1995 Chagnon told *Scientific American* editor John Horgan that because male aggression is esteemed in Yanomamö culture, aggression as a human trait is highly malleable and culturally influenced, an observation that might have been made by Stephen Jay Gould, considered by most sociobiologists to be Satan incarnate. "Steve Gould and I probably agree on a lot of things," Chagnon surprisingly concluded.[28]

Even when he is talking about the Yanomamö casually and not for publication, Chagnon carefully nuances and contextualizes everything he says. For example, at the Skeptics Society 1996 Caltech conference on evolutionary psychology Chagnon delivered a fact-packed lecture mixing anecdotes and data, including the graphs from his now-famous *Science* article revealing the positive correlation between levels of violence among Yanomamö men and their corresponding number of wives and offspring. "Here are the 'Satanic Verses' that I committed in anthropology," Chagnon joked, as he reviewed his data:

> I didn't intend for this correlation to pop out, but when I discovered it, it did not surprise me. If you take men who are in the same age category and divide them by those who have killed other men (unokais) and those who have not killed other men (non-unokais), in every age category unokais had more offspring. In fact, unokais averaged 4.91 children versus 1.59 for non-unokais. The reason is clear in the data on the number of wives: unokais averaged 1.63 wives versus 0.63 for non-unokais. This was an unacceptable finding for those who hold the ideal view of the Noble savage. "Here's Chagnon saying that war has something good in it." I never said any such thing. I merely pointed out that in the Yanomamö society, just like in our own and other societies, people who are successful and good warriors, who defend the folks back home, are showered with praise and rewards. In our own culture, for example, draft dodgers

are considered a shame. Being a successful warrior has social rewards in all cultures. The Yanomamö warriors do not get medals and media. They get more wives.[29]

And despite the mountains of data Chagnon has accumulated on Yanomamö aggression, he was careful to note throughout his lecture the many other behaviors and emotions expressed by the Yanomamö: "When I called the Yanomamö the 'fierce people,' I did not mean they were fierce all the time. Their family life is very tranquil. Even though they have high mortality rates due to violence and aggression and competition is very high, they are not sweating fiercely, eating fiercely, belching fiercely, etc. They do kiss their kids and are quite pleasant people."[30]

Even in the question-and-answer period, when given the opportunity to make his case for an extreme sociobiological view of humans to an obviously receptive audience dominated by older males who encouraged such answers with leading questions, Chagnon gently but firmly demurred. One gentleman inquired whether Chagnon thought that his data implied there might be genes for violence that are passed down to future generations by the unokais, and whether this implied that perhaps all human violence is innate. Chagnon unhesitatingly answered in the negative: "No, I do not think violence is that directly connected to specific genes, although there is undoubtedly a biological substrate underlying violence. Violence is a facultative trait. You have to look at the environmental cues to see what touches it off. Because they are an inbred population we can expect that Yanomamö genes are different from other populations, but I do not think that they are any different genetically from other populations in terms of violence."[31]

In light of his data on warriors who are rewarded with more wives, one questioner wondered what happens to the men who get no wives, and if this means that the Yanomamö are polygamous. Chagnon explained that, indeed, some Yanomamö men have no wives and that it is often they who are the causes of violence as they either resort to rape or stir up trouble with men who have more than one wife. But he added an important proviso that indicates, once again, Chagnon's sensitivity to the nuances and complexities within all cultures, and the danger of gross generalizations: "Anthropologists tend to pigeonhole societies as monogamous or polygamous or polyandrous, as if these are three different kinds of societies. In

fact, you have to look at marriage as a life-historical process in all societies. There are, for example, cases of monogamy in Yanomamö society. In
fact, monogamy is the most common type of marriage. But there are also
polyandrous families where one woman marries two men, who tend to
be brothers. There are, in fact, examples of all three types of marriage
arrangements in Yanomamö culture."[32]

Even at work Chagnon refrains from oversimplifying his research. I
asked anthropologist Donald Symons, Chagnon's colleague from the University of California, Santa Barbara, about the accusations that Chagnon
is a sociobiological ideologue bent on painting a portrait of humanity as
self-centered, competitive, and violent. Symons replied: "You know, it's
interesting that people make such charges against Nap, because when you
ask him about this or that aspect of the Yanomamö he never just offers
some simple opinion. He'll says things like 'I think I can get that data for
you,' or 'let me check that and get back to you.'" So then why does
Chagnon seem to have so many enemies, I inquired? "Well, sometimes he
responds to his critics in a belligerent manner that is off-putting to many
people. His initial defense is typically ad hominem, where he will call his
critics Marxists or Rousseauian idealists. That's not the way to defend
against charges, which should be answered point by point."[33]

That is, in fact, what is being done by a cadre of Chagnon defenders,
who have compiled an impressive literature of point-by-point refutations
of Tierney's accusations. Even Chagnon was taken by surprise. "I've
received a number of e-mails from people identifying themselves on the
academic left, who made it clear that while they disagree with me on a
number of theoretical points they do not want anything to do with Tierney
or his book."[34]

Spin-Doctoring Science

In politics, spin-doctoring is the art of interpreting words and actions in a
light favorable to one's position or cause. Spin-doctoring is openly practiced in politics and spin doctors have become star players on politicians'
teams (there was even a television show called *Spin City* that revolves
around a spin doctor played by Michael J. Fox). While spin-doctoring has
and does go on in science, ideally we strive for objectivity and we hope

that peer review and the other checks and balances that are part of the self-correcting nature of science keep it at a tolerable minimum.

What we are witnessing in this latest battle in the anthropology wars is journalistic spin-doctoring of what is, for the most part, solid science. In carefully reading Good's *Into the Heart* and Chagnon's *Yanomamö* back to back over the course of several days of intense study, I found myself continually wondering how Tierney could possibly have read both books and come away with the impressions he did, unless this was a clear-cut case of spin-doctoring. The same descriptions of violence, aggression, and especially rape are present in both books; it all depends on the "spin" one puts on the data. For example, Good writes:

> I got increasingly upset about Chagnon's "Fierce People" portrayal. The man had clearly taken one aspect of Yanomami behavior out of context and in so doing had sensationalized it. In the process he had stigmatized these remarkable people as brutish and hateful. I wasn't fooling myself into thinking that the Yanomami were some kind of Shangri-la race, all peace and light. Far, far from it. They were a volatile, emotional people, capable of behavior we would consider barbaric.[35]

Well, if the Yanomamö are really "barbaric," then why is it sensationalistic to call them "brutish"? It all depends on the spin.

Into the Heart is a page-turner because the very features of Yanomamö culture that Chagnon's critics claim he overemphasizes are, in fact, present in spades in every chapter of Good's gripping tale. As Chagnon's graduate student, Good emersed himself in Yanomamöland but in time found himself falling in love with a beautiful young Yanomamö girl named Yarima. (Columbia Pictures bought the rights to produce a dramatic film based on the book, and Good even received a phone call from the actor Richard Gere, who was interested in playing him. That deal has since fallen through and others have shown interest in a film deal, but nothing has come of it to date. Good has avoided commenting publicly about the Tierney–Chagnon controversy, he said, because he doesn't want his half-Yanomamö children to become the focus of media attention.[36])

As the years passed and he had a falling out first with Chagnon and then with the world-renowned ethologist Irenäus Eibl-Eibesfeldt, Good became emotionally distraught over leaving Yarima alone when he was forced to return to Caracas to renew his permit, or when he was to return to the United States or Germany to attend conferences or work on his doctoral dissertation. Why? When Yarima came of age (defined as first menses in Yanomamöland), she and Good began living together and consummated their "marriage." (Yanomamö do not have a marriage ceremony per se; instead a couple, usually the man, declares that they are married and the two begin living together.) Good's problem was that he was all too aware of what Yanomamö men are really like:

> They will grab a woman while she is out gathering and rape her. They don't consider it a crime or a horrendously antisocial thing to do. It is simply what happens. It's standard behavior. In such a small, enclosed community this (together with affairs) is the only way unmarried men have of getting sex.[37]

Good's worries were justified and the universal emotion of jealousy was no more attenuated in this highly civilized, educated man than it was in any of the people he was studying to earn his Ph.D. In short, Good was on an emotional roller coaster from which he could not extricate himself.

> I felt the tension, and I tried to deal with it. I wanted to think that Yarima would be faithful to me. But I knew the limits of any woman's faithfulness here. Fidelity in Yanomami land is not considered a standard of any sort, let alone a moral principle. Here it is every man for himself. Stealing, rape, even killing—these acts aren't measured by some moral standard. They aren't thought of in terms of proper or improper social behavior. Here everyone does what he can and everyone defends his own rights. A man gets up and screams and berates someone for stealing plantains from his section of the garden, then he'll go and do exactly the same thing. I protect myself, you protect yourself. You try something and I catch you, I'll stop you.[38]

Many antisocial behaviors, such as theft, are kept at a minimum through such social constraints as shunning, or personal constraints as fear of violence and retaliation. But, as Good explains, "sex is a different story."

> The sex drive demands an outlet, especially with the young men. It cannot be stopped. Thus the personal and social constraints have

less force, they're more readily disregarded. As a result, a woman often has no choice. And if a woman is raped, she will not tell her husband, because she knows that her husband will beat her, or worse. In most cases the husband will become extremely angry, at both his wife and at the man who has raped her. But his anger will most likely not have the intensity or duration to provoke a village-shattering conflict, unless perhaps his wife is young and has not had a child yet. In that case the husband might find he cannot tolerate it; he might lose control utterly and embark on violent action. He badly wants to at least get his family started himself, rather than have someone else make her pregnant.[39]

How different is Good's analysis from Chagnon's description of a raid triggered by a desire to seek revenge for an earlier killing that also included the abduction of women.

Generally, however, the desire to abduct women does not lead to the initiation of hostilities between groups that have had no history of mutual raiding in the past. New wars usually develop when charges of sorcery are leveled against the members of a different group. Once raiding has begun between two villages, however, the raiders all hope to acquire women if the circumstances are such that they can flee without being discovered. If they catch a man and his wife at some distance from the village, they will more than likely take the woman after they kill her husband. A captured woman is raped by all the men in the raiding party and, later, by the men in the village who wish to do so but did not participate in the raid. She is then given to one of the men as a wife.[40]

It's all in how one spin-doctors the data. Chagnon carefully parcels it into contextualized units as quantitative data. Good uses it anecdotally as part of a literary narrative, and Tierney uses Good to weave it into an indictment. For example, Tierney accuses Chagnon of acting "fiercely" around the Yanomamö, dressing up like them, using their drugs, pounding his chest, and screaming. Tierney gleefully explains how "Chagnon suddenly went from being an impoverished Ph.D. student at the bottom of the totem pole to being a figure of preternatural power," then quotes Chagnon's own description of how he immersed himself into the Yano-mamö culture.

The village I'm living in really thinks I am the be-all and the end-all. I broke the final ice with them by participating in their dancing and singing one night. That really impressed them. They want to take me all over Waicaland to show me off. Their whole attitude toward me changed dramatically. Unfortunately, they want me to dance all the time now. You should have seen me in my feathers and loincloth![41]

Contrast this passage, quoted by Tierney to incriminate Chagnon, with Good's explanation of why he did nearly the same thing as Chagnon did in order to be accepted into Yanomamö culture.

What people, even some anthropologists, do not understand is how truly different it is to live with people whose conception of morality, laws, restrictions, controls differs so radically from ours. If you don't protect yourself, if you don't defend yourself, if you don't demand respect—you don't survive. It's as simple as that. If you act down here as you would up there, you'll be so intimidated, so worked over, you'll be running out of here. And lots of guys have been.[42]

Despite these rather brutish descriptions of the Yanomamö, Good concludes: "The point is that it's what you want to see, it's what you are drawn to write about. And that's supposedly anthropology. Chagnon made them out to be warring, fighting, belligerent people, confrontations, showdowns, stealing women, raping them, cutting off their ears. That may be his image of the Yanomami; it's certainly not mine."[43]

This is spin-doctoring. The entire second half of Good's book is a spellbinding narrative about how Good spent most of his time warding off men from his wife who, despite his best efforts, was gang-raped, beaten, had an earlobe torn off, and was stolen by a man while Good was away renewing his permit. And Good admits that this "was more or less standard conduct. Men will threaten, and they'll carry out their threats, too. They'll shoot a woman for not going with them. I know of more than one woman who has been killed for rejecting advances made under threat. What usually happens is that she goes along with it. There isn't any choice. You go and make the best of it."[44]

Who is the Hobbesian anthropologist? As an outsider with no relationship to any of the players in this anthropological drama, and no commitment to any theoretical position within the science, I fail to see any difference between Chagnon's description of the Yanomamö and Good's. Tierney has merely spin-doctored them toward his cause. The photographs in figures 5.1 and 5.2 visually illustrate these descriptions.

Science as a Candle in the Dark

The psychology and sociology of science is interesting and important in understanding the history and development of scientific theories and ideas, but the bottom-line question here is this: Did Chagnon get the science right? Some anthropologists question the level of violence reported by Chagnon, claiming that they have recorded different (and often lower) rates in other areas of Yanomamöland. Good, for example, is quoted by Tierney as saying: "In my opinion, the Fierce People is the biggest misnomer in the history of anthropology." Did he mean that, I asked him?

> All along I have felt that Chagnon has not represented the Yanomamö accurately. I feel he might have slanted, or even cooked, some of his data. In over a dozen villages in the course of a dozen years of research, I never saw what Chagnon reported that he saw in terms of the violence both within and between villages. I have said from the beginning that (1) Chagnon did bad fieldwork, or (2) there was something wrong with the villages Chagnon studies, or (3) Chagnon was in a "hot" area with a lot of activity whose disruptive forces led to higher levels of violence.[45]

What about Chagnon's data, I wondered? It strikes me as heavily quantitative and easily checkable. "Yeah, well, let me tell you about anthropological 'data,'" Good retorted. "An anthropologist goes into the field for fifteen months, comes out, and tells the world what the people he has studied are like. No one was there with him. How can those observations be checked? This is a serious problem in anthropology. I'm sure that ninety percent of anthropologists are doing good fieldwork, but in my opinion there are problems with both Chagnon's and Lizot's data." Perhaps he and Chagnon had studied different Yanomamö villages and this might account for their differing conclusions? "I went to several of the same villages as Chagnon, and I just didn't see what he saw."[46]

Figure 5.1. Yanomamö men duel over a woman. Despite Patrick Tierney's claims that Napoleon Chagnon exaggerated the level of aggression and rape among the Yanomamö, both behaviors have been documented.

Figure 5.2. Yanomamö war scar. Many Yanomamö men have deep scars on their heads from such battles.

Considering his reputation I half expected Chagnon to explode over the phone when I queried him about these charges. Not only did he respond with dispassionate coolness, he had nothing critical at all to say about Good, commenting only that it is entirely possible that he and his former student did see different behaviors: "Good spent much of his time trekking with the Yanomamö, going on hunting trips outside of the village. If a village contains, say, one hundred and fifty people in a complex web of relations, but you are spending most of your time with just a dozen or so away from the village, of course you are going to make different observations."[47]

In *Yanomamö* Chagnon notes that such variation in violence observed by different scientists can be accounted for by a concatenation of intervening variables, such as geography, ecology, population size, resources, and especially the contingent history of each group where "the lesson is that past events and history must be understood to comprehend the current observable patterns. As the Roman poet Lucretius mused, nothing yet from nothing ever came."[48]

After reading through the literature and interviewing many on all sides of this issue, my conclusion is that Chagnon's view of the Yanomamö— while in the service of some view (a Darwinian one)—is fundamentally supported by the available evidence. His data and interpretations are corroborated by many other anthropologists who have studied the Yanomamö. Even at their "fiercest," however, the Yanomamö are not so different from many other peoples around the globe (recall Captain Bligh's numerous violent encounters with Polynesians, and Captain Cook's murder at the hands of Hawaiian natives), even when studied by tender-minded, non-fierce scientists. (I do find it ironic, however, that in their attempts to portray the Yanomamö as Rousseauian noble savages—in defense of a view of human nature as basically benign and infinitely flexible—a few anthropologists have proven themselves to be fierce and persistent warriors in their battles with Chagnon.)

Evolutionary biologist Jared Diamond, for example, told me that he found the role of warfare among the peoples of New Guinea that he has studied over the past thirty years quite similar to Chagnon's depiction of the role of warfare among the Yanomamö.[49] And, judging by the latest archaeological research presented in such books as Arthur Ferrill's *The Origins of War* and Lawrence Keeley's *War Before Civilization*, Yanomamö violence and warfare are no more extreme than those of our paleolithic ancestors who, around the world and throughout the past twenty thousand years, appear to have brutally butchered one another with all too frequent abandon.[50] Finally, if the past five thousand years of recorded human history is any measure of a species' "fierceness," the Yanomamö have got nothing on either Western or Eastern "civilization," whose record includes the murder of hundreds of millions of people.

Homo sapiens in general, like the Yanomamö in particular, are the erotic-fierce people, making love and war far too frequently for our own good as both overpopulation and conflict threaten our very existence. Just as science has been our candle in the dark illuminating our path into the heart of human nature, science is our greatest hope for the future, showing us how best we can utilize our natures to ensure our survival.

6

Psyched Up, Psyched Out

Can Science Determine if
Sports Psychology Works?

QUESTION: HOW MANY PSYCHOLOGISTS does it take to change a lightbulb? Answer: Only one—but the lightbulb has to really want to change. In this lighthearted joke there is much insight into the efficaciousness of psychological techniques to modify human thought and behavior, because on a fundamental level the willingness of the subject to participate in the modifying sets the stage for everything else that follows.

Although I was trained as an experimental psychologist, I did not become interested in understanding the range and power of psychological techniques to enhance athletic performance until 1981 when I began preparations to compete in the first annual three-thousand-mile nonstop transcontinental bicycle race called Race Across America (see figure 6.1). Being an experimental psychologist and curious fellow, I thought it reasonable to try any and all techniques I could find to prepare my mind for the pain and pressures of what *Outside* magazine called "the world's toughest race."

In addition to riding five hundred miles a week and subjecting my body to such "treatments" as chiropractic, Rolfing, mud baths, colonics, megavitamin therapy, iridology, food allergy tests, and electrical stimulation, I listened to motivation tapes, to specially designed music tapes, and even

Figure 6.1. Really psyched. The author climbing Loveland Pass, twelve thousand feet up in the Colorado Rockies, on day three of the three-thousand-mile nonstop transcontinental bicycle Race Across America.

to subliminal tapes (although being subliminal there isn't much to hear). I meditated. I chanted. I attended seminars from an Oregon-based healing guru named Jack Schwarz who taught us "voluntary controls of internal states." I contacted an old associate of mine with whom I went to graduate school, Gina Kuras, who was now a practicing hypnotherapist. Gina taught me self-hypnosis, from which I would learn to control pain, overcome motivational lows, maintain psychological highs, and stay focused. I got so good at going deep into a hypnotic trance that when ABC's *Wide World of Sports* came to my home to film a session, Gina had a fright when she could not immediately bring me back to consciousness.

Anecdotes Do Not Make a Science

Did all this vintage 1980s New Age fiddle-faddle work? One answer to this question comes from one of the apostles of the motivation movement, Mark Victor Hansen (now famous for his *Chicken Soup for the Soul* book series), whom I met in 1980. "This stuff works when you work it," he would chant. On one level Mark is right. As with the lightbulb, to change you have to want to change and then make it happen yourself. Like ever-popular fad diets, it matters less which one you are on and more that you are doing something—anything—about your weight. To that end, diets are really a form of behavioral, not caloric, modification. The point is to be vigilant and focused, thinking about the problem and trying different solutions.

But a deeper and more important question is this: Can we say scientifically that something works? That question is far more complicated because so many of these self-help techniques are based on anecdotal evidence, and as my social science colleague Frank Sulloway likes to point out: "Anecdotes do not make a science. Ten anecdotes are no better than one, and a hundred anecdotes are no better than ten."

Anecdotes are useful for helping focus research on certain problems to be solved, but the reason they do not make a science is because, without controlled comparisons, there is no way to know if the effect that was observed was due to chance or to the technique. Did you win the race because of the meditation, or was it because you had a deep sleep, or a good meal, or new equipment, or better training? Even if ten athletes who applied a certain procedure before an event performed better, without a control comparison group there is no way to know if their improvement was due to that or to some other set of variables they shared, or to statistical chance. And when we say that an athlete performed "better," better than what? Better than ever? Better than yesterday? Better than average? Conducting a scientific evaluation of the effectiveness of psychological procedures on athletic performance is a messy and complicated process. But science is the best method we have for understanding causality, so apply it we must.

The Science of Sports Psychology

What is the evidence for the validity of sports psychology as an applied science? Those who are not into sports will be surprised to learn that there is a sizable body of scientific literature on the subject, with practitioners teaching courses, attending conferences sponsored by professional societies (e.g., Association for the Advancement of Applied Sport Psychology), publishing in professional journals (e.g., *Journal of Sport and Exercise Psychology*), and authoring books specializing in the scientific study of how psychological factors influence physical performance, and how physical activity affects psychological states.

Apropos my connection to the field, sports psychology began in the 1890s when Indiana University psychologist Norman Triplett, an avid cyclist, conducted a series of studies to determine why cyclists ride faster in groups than when alone. Triplett discovered that the presence of other competitors and/or spectators motivates athletes to higher levels of performance, and he corroborated this theory on other activities, such as with children who reeled in more line on a fishing pole while in the presence of other children involved in the same activity. As sports became professionalized in the early twentieth century, psychologists realized that psychological states could influence physical performance in profoundly significant ways. Since then the field has paralleled the trends of psychology in general, applying behavioral models (e.g., how rewards and punishment shape performance), psychophysiological models (e.g., the relationship between heart rate and brain-wave activity and performance), and cognitive-behavioral models (e.g., the connection between self-confidence and anxiety on performance). And as in its parent field, the goal of sports psychology is to apply such theoretical models to real-world situations for the purpose of understanding, predicting, and controlling the thought and behavior of athletes.

Consider the above example of cyclists riding faster in a pack. Controlling for the well-known aerodynamic effect called drafting, studies show that a cyclist will ride faster even when another cyclist is just riding alongside or even behind, and on average cyclists will race faster against a competitor than against the clock. Why? One reason is what is known as social facilitation, a theory with broad application to many social situations where individual behavior is shaped by the presence and motivation of a group (think of mass rallies and rock concerts). But social facilitation is just a term. What is actually going on inside the brain and body? Utilizing components from all three interpretive models we see that competition provides the promise of positive (and threat of negative) reinforcements, stimulates an increase in physiological activity and arousal, and locks the athlete into an autocatalytic feedback loop between performance expectations and actual outcomes. This constant feedback causes competitors to push each other to the upper limits of their physical capabilities.

Mr. Clutch vs. Mr. Choke

As in all psychological theories, there are intervening variables that qualify the effect. For many athletes, competition and crowds cause an increase in anxiety that hampers performance. They crumble under the pressure of fan expectations. They choke under the lights of the crowded stadium. The acceleration of heart rate and adrenaline that accentuates one competitor's drive to win attenuates another's. Basketballs that go swish in practice go clunk in the game. Aces on the practice court turn into double faults at center court. Crushed homers during batting practice become whiffs as soon as the ump cries, "Play ball."

Sports psychologists offer several explanations for this variance. Situational states affect personality traits, and these traits vary, with some athletes far more at ease under pressure (think of Reggie Jackson as "Mr. October" or Jerry West as "Mr. Clutch"), while others falter (think of Bill Buckner's infamous through-the-legs error that squelched the Boston Red Sox's best chance for a World Series victory or Scott Norwood's muffed field goal in the closing seconds of the Buffalo Bills' only opportunity for a Super Bowl ring). Personality research on elite athletes shows that they are high in stimulus seeking, risk taking, competitiveness, self-confidence, expectation for success, and the ability to regulate stress. They practice a lot, come prepared with a contingency plan for changes in the competition, stay focused on the event, block out distracting stimuli, rehearse their task mentally before the game, follow their own plans and not those of the competitor, are not flustered by unexpected events, learn from mistakes, and never give up. Personality matters.

The Hot-Streak Myth

Such psychological influences also depend on the complexity of the task, with social facilitation, for example, more beneficial for simple and well-practiced skills but less helpful for difficult tasks that are less rehearsed. Pedaling a bicycle or lifting a barbell is not in the same league of complexity as putting a golf ball or executing a gymnastics routine. The hundred thousand screaming fans lining the final kilometers of a leg-burning climb up the French Alps in the Tour de France that might catapult a cyclist onto the winner's podium could cause a golfer to knock a chip into the sand trap or a gymnast to do a face-plant into the mat. Context counts. The anthropologist Bronislaw Malinowski, for example, discovered that the level of superstition among Trobriand Island fishermen depended on the level of uncertainty of the outcome—the farther out to sea they went, the more complex their superstitious rituals became. If we think of fishing as a sport, we see a striking parallel in baseball where batters, the best of whom fail nearly seven times out of every ten trips to the plate, are notorious for their superstitious rituals, whereas fielders, typically successful nine out of every ten times a ball is hit to them, have correspondingly fewer superstitions. And remember that this difference is expressed in the same players!

Also affecting athletic performance is physiological arousal, depicted in the infamous "inverted-U" diagram—the two feet of the inverted U are firmly planted on the low end of the performance gradient while the neck of the U represents peak performance. Levels of arousal too low or too high are deleterious, but a medium amount generates what is needed for optimal operation. The principle also applies to the so-called home court advantage. Anecdotally, we all "know" that competitors have an advantage when playing at home, and teams strive all season to finish with the best record in order to get it. But research qualifies the effect. On average and in the long run, football and baseball teams do slightly better at their own stadiums than at their competitors', and basketball and hockey teams do significantly better at home than away (the smaller arenas of the latter presumably enhance social facilitation). But here we might think of regular season play as falling into the idealized middle of the inverted U, with preseason on the low arousal end and postseason on the high. For example, one study showed that in the baseball World Series from 1924 to 1982, in series that went five games or more, the home team won 60 percent of the first two games but only 40 percent of the remaining games. Interestingly, in the twenty-six series that went to a nail-biting seventh game, the home team came away empty-handed 62 percent of the time. Since 1983, however, the trend has shifted somewhat. Between 1983 and 1999 the home team won 54 percent of the first two games, but in series that went five games or more the home team won 62 percent of the remaining games, and 80 percent of the deciding seventh game. It is possible that in some instances overly zealous fans become fanatics (from whence the term comes) in the final stretch, driving their charges into a frenzied state of unrealistic expectations that stymies performance. But clearly there are teams that thrive on such pressure and respond accordingly.

Science has also illuminated (and in one case debunked) another aspect of the psychology of sport known technically as Zones of Optimal Function, but colloquially called peak performance, or flow. This is "runner's high." It is one of those fuzzy concepts athletes talk about in equally fuzzy expressions such as being "in sync," "floating," "letting go," "playing in a trance," "in the cocoon," "in the groove," or "going on autopilot." Psychologists describe it with such adjectives as relaxed, optimistic, focused, energized, aware, absorbed, and controlled. It is a matching of skills with the challenge at hand. And it is called peak performance because, supposedly, the athlete's performance hits a peak in this state of flow. The golf ball drops into the cup instead of skirting the edge. The bat hits the ball with that crisp crack and it always seems to fall where they ain't. Basketballs drop one after another, swish, swish, swish. When you're hot, you're hot.

Maybe not. Streaks in sports can be subjected to the tests of statisticians who consider the probability of such trends in any given task. Intuitively we believe hot streaks are real, and everyone from casino operators to sports bookies counts on us to act on this belief. But in a fascinating 1985 study of "hot hands" in basketball, Thomas Gilovich, Richard Vallone, and Amos Tversky analyzed every basket shot by the Philadelphia 76ers for an entire season and discovered that the probability of a player hitting a second shot did not increase following an initial successful basket, beyond what one would expect by chance and the average shooting percentage of the player. In fact, what they found is so counterintuitive that it is jarring to the sensibilities: the number of streaks, or successful baskets in sequence, did not exceed the predictions of a statistical coin-flip model. That is, if you conduct a coin-flipping experiment and record heads or tails, you will shortly encounter streaks. How many streaks and how long? On average and in the long run, you will flip five heads or tails in a row once in every thirty-two sequences of five tosses. Since we are dealing with professional basketball players instead of coins, however, adjustments in the formula had to be made. If a player's shooting percentage was 60 percent, for example, we would expect, by chance, that he will sink six baskets in a row once for every twenty sequences of six shots attempted. What Gilovich, Vallone, and Tversky found was that there were no shooting sequences beyond what was expected by chance. Players may feel "hot," and when they have games that fall into the high range of chance expectations they feel "in flow," but science shows that nothing happens beyond what probability says should happen. (The exception is Joe DiMaggio's fifty-six-game hitting streak, a feat so many standard deviations away from the mean that, in the words of the scientists who calculated its probability, Ed Purcell and Stephen Jay Gould, it "should not have happened at all" and ranks as perhaps "the greatest achievement in modern sports." Every once in a while individual greatness defies scientific theory.)

The Art of Sports Psychology

The fortunes and failures of sports psychology match those of the social sciences in general: we are much better at understanding behavior than we are in predicting or controlling it. It is one thing to model all the variables that cause some athletes to triumph and others to flounder, and even apply multiple analyses of variance to groups of athletes and estimate the percentage of influence each variable has, on average and in the long run, on outcomes. It is quite another to make specific predictions of which athletes will step up on the winner's podium at the end of the day, and virtually impossible to turn Willy Whiff into Mark McGwire or Andy Airball into Michael Jordan. Here we enter the murky world of performance enhancement and sports counseling. As in the general field where experimental psychology is the science and clinical psychology the art, this is the art of sports psychology.

One of the most commonly practiced and effective sports psychology intervention techniques is called imagery training, or visualization, where you "see" (although you should also "feel") yourself executing the physical sequence of your sport. We have all seen Olympic downhill skiers minutes before their run standing in place with their eyes closed and body gyrating through the imaginary course. Gymnasts and ice-skaters are also big on imagery, but even cyclists can benefit—witness Lance Armstrong's remarkable 1999 Tour de France victory to which he attributed his success, in part, to the fact that he rode every mountain stage ahead of time so that during the race itself he could imagine what was coming and execute his preplanned attacks. And countless experiments on imagery show time and again that groups who receive physical and imagery training on a novel task do better than groups who receive physical training only. Imagery training fits well into cognitive-behavioral models of sports psychology.

Nevertheless, failures of imagery-trained athletes are legion in sports. We hear about the Lance Armstrong stories because we love winners and want to emulate what they did. We simply never hear about all those cyclists who also rode the Tour stages ahead of time but finished in the middle of the pack, or the visualizing downhill skiers who crashed, or the imagining gymnasts who flopped. Even the most enthusiastic supporters of imagery training for peak performance (or any other clinical method) caution that there are numerous mediating variables that can prevent an individual from benefiting from the technique.

For example, sports psychologists Daniel Gould and Nicole Damarjian caution that imagery takes time to develop as a skill and you must practice it on a regular basis, use all your senses in the imagery process, apply both internal and external perspectives, and utilize relaxation techniques in conjunction with imagery, as well as video- and audiotapes and logs to record the process and progress. They conclude that imagery is not for everyone and that "it is important to remember that imagery is like any physical skill in that it requires systematic practice to develop and refine. Individual athletes will differ in their ability to image and, therefore, must be encouraged to remain patient. Imagery is not a magical cure for performance woes. It is, however, an effective tool that—when combined with practice and commitment—can help athletes reach their personal and athletic potentials." In other words, they seem to be saying that this stuff works when you work it. But what does that mean?

Flooded with Flapdoodle

To determine if a psychological technique works, we might evaluate it by two standards: criterion 1 (does it work for an individual?) and criterion 2 (does it work for all individuals?). For criterion 1, to the athlete who wins the big meet or the gold medal, whatever he or she did "worked." Like the depressed person who leaves therapy happy, it does not matter what scientists think of the therapy, because the therapy worked in the sense of doing something that resulted in a positive outcome. Most practicing therapists accept criterion 1 as good enough.

But to the extent that psychology (including sports psychology) is a science that would like to be able to claim, at least provisionally, that a clinical technique is something more than just talking to a good friend or watching Marx brothers movies (or even doing nothing at all), criterion 2 becomes important. Here we face a problem that hangs like an albatross around the neck of the entire profession of clinical psychology and that has drawn, if you will, an iron curtain between clinicians and experimentalists—there is very little experimental evidence that therapeutic techniques really work in the sense of criterion 2. I do not go as far as psychiatrist Thomas Szasz in his claim that mental illnesses are all socially constructed, nor do I accept all of clinical psychologist Tana Dineen's arguments that the "psychology industry" is "manufacturing victims" in order to feed its ever-growing economic juggernaut. But Szasz and especially Dineen (and social psychologist Carol Tavris in a much more reasoned approach) have injected a badly needed dose of skepticism into a field flooded with flapdoodle.

Sports psychology has its fair share of the same, as I discovered in my journeys as a cyclist and a scientist, and both practitioners and participants would be well advised to step back and ask themselves if criterion 1 (the art) is good enough and, if not, how criterion 2 (the science) can become attainable. Did all the psychological exercises I tried "work" for me in the Race Across America? It is impossible to say, because I was a subject pool of one and there were no controls. When I wanted them to work, it seemed like they did, and maybe that's good enough. Yet I cannot help but wonder if a few more hours in the training saddle every day might have made a bigger difference. Sports can be psychological, but they are first and foremost physical. Although body and mind are integrated, I would caution not to put mind above body.

7

Shadowlands

Science and Spirit in Life and Death

SIXTY THOUSAND YEARS AGO, in a cave 132 feet deep cut into the Zagros Mountains of northern Iraq, 250 miles north of Baghdad at a site called Shanidar, the body of a Neanderthal man was carefully buried in a cave, on a bed of evergreen boughs, on his left side, head to the south, facing west, and covered in flowers, so identified through microscopic analysis of the surviving pollens. Already in the grave were an infant and two women. The flowers were from eight different species and the arrangement was not accidental. There was a purpose to the burial process. It is the earliest memorial celebration of life and mourning of death of which we know.

Now that Neanderthals are extinct, we are the only species who is aware of its own mortality. Death is an inescapable end to life. Every organism that has ever lived has died. There are no exceptions. Behind every one of the 6.2 billion people now living lie seventeen others in the ground, for 106 billion, according to the demographer Carl Haub, is the total number of humans who have ever lived. Our future is sealed by our past.

Thus, we are faced with the existential question that has haunted everyone who has thought about this uncomfortable fact of life: Why are we here? People throughout the ages and around the world, in all cultures and communities, have devised a remarkable variety of answers to this question. Indeed, anthropologists estimate that over the past ten thousand years humans have created roughly ten thousand different religions, the wellspring of which may be found in the answers they have offered to that soul-jarring question: Why are we here?

I started thinking hard about this question in 1992, when my mother started acting strange—she was confused a lot of the time, disoriented, and emotionally unstable. Since my folks had just been dealt a significant financial blow, I thought that perhaps this triggered her erratic behavior because she grew up in the Depression and always feared that she would be destitute again. I spent countless hours talking to her, as did her sister, her friends, and my dad. I escorted her on long walks in hopes that some physical exercise might release some mood-elevating endorphins.

Months passed with no sign of improvement, so we took her to a psychiatrist in Pasadena, California, who, after half an hour, diagnosed depression and, of course, prescribed an antidepressant. I was skeptical. My mom was weird, not depressed. I asked for a second opinion, perhaps one from a neurologist. The psychiatrist thought this a good idea . . . after trying the antidepressants for several months. Insisting that we get a second opinion *that day*, I asked for the referral number, walked over to her phone, dialed the number, and announced, "This is Dr. Shermer over at Dr. Smith's office. We have a patient we need to get in today." To my utter amazement, this minor deception worked! (I'm a Ph.D., not an M.D.)

I'll never forget that appointment, or when it was—October 1992. I remember because it was in the middle of the presidential debates between George Bush and Bill Clinton, and the neurologist asked my mom who was running for president. She didn't know. Minutes later they were wheeling her in for a CAT scan. We returned after lunch to see the results. Since I was still "Dr." Shermer, I was taken into the inner sanctum to see the results. I nearly fainted at what I saw—a massive tumor filled the front half of the left side of my mom's skull, smashing her neurons against the back of her brain case. Through a shaky voice I explained to her what was going on, and that the solution would be major brain surgery, which can be extremely risky. But because her brain was squashed and she was not thinking clearly, she just shrugged and quietly acknowledged, in a small but sweet voice, "Okay."

My mom's tumor was a meningioma (originating on the meninges, the protective lining of the brain), which are far more common in women than men—the ratio ranges from 1.4:1 to 2.8:1—and cluster, curiously, in Los

Angeles County, where we live. (Tragic irony: the psychiatrist who diagnosed my mom with depression was herself later diagnosed with the same type of brain tumor.) Fortunately, most meningioma tumors are successfully resected. Indeed, within days of the surgery her brain filled back into the hole and my mom was back to her bright and cheery self—what a remarkably recuperative and pliable organ is the brain.

Unfortunately, my mom might be another data point in support of Harvard Medical School surgeon Judah Folkman's controversial angiogenesis theory of cancer, which he developed in the 1960s while conducting research for the U.S. Navy. Folkman noticed that tumors planted in isolated organs in glass chambers all stopped growing at a certain size. Their size range should have exhibited the classic bell curve, so he deduced that it was a lack of blood vessels in the tumors that prevented their continued development. In a 1971 paper in the *New England Journal of Medicine*, Folkman outlined his theory in four stages: (1) blood vessels in tumors are new, which means the tumors have to recruit them; (2) the tumor recruiting agent is TAF, or tumor angiogenesis factor; (3) the tumor secretes the TAF, which draws blood vessels to it; (4) once the production of TAF is terminated, the tumor ceases to grow.

By the 1980s, Folkman had discovered a number of angiogenesis inhibitors, including interferon and platelet factor 4, and in the late 1990s endostatin and angiostatin were found to be highly effective retardants of mice tumors. Today about twenty different compounds are being tested for their antiangiogenesis capacity, including Avastin and Erbutix, but thus far none of the angiogenesis inhibitors has been consistently effective in attenuating tumor growth in humans.

Most interesting to me was Folkman's contention that tumors secrete their own angiogenesis inhibitors that prevent other tumors from initiating angiogenesis programs. Remove the dominant tumor, however, and you eliminate the angiogenesis inhibitors, thus allowing other dormant tumors to spring to life. Sure enough, within months of my mom's initial brain surgery, two new tumors appeared. Fortunately they were within reach of the surgeon's scalpel, but their removal in March 1993 apparently allowed other quiescent tumors to awaken, leading to additional craniotomies in July 1993 and September 1994.

In my mind I had always thought of "tumors" as somehow not as bad as "cancer." In fact, tumors may be benign or malignant, depending on how encapsulated and contained they are versus tentacled and spreading. After four craniotomies in two years my mom's tumor got upgraded to an "invasive malignant meningioma." It was even more invasive than her physicians realized. In 1999, she unexpectedly developed two tumors in her lungs. Could it be that the invasive malignant meningioma had migrated out of her brain? Impossible, we were told. By Folkman's theory, however, this is possible, because the plethora of micro blood vessels that infuse a tumor increase the probability of single cells escaping the encapsulated tumorous mass and invading the general circulatory system. Not likely, was the response we received to this suggestion, since there is no known case of such a migration by a meningioma tumor, benign or malignant.

Here is yet another insight into the nature of the scientific process. At a dinner party one evening, hosted by my friend the social psychologist Carol Tavris, I was recounting my mom's plight, including this latest turn of events. Quite by chance, also dining with us was Dr. Avrum Bluming, a practicing medical oncologist who also specialized in the dissemination of medical information via the Internet, having founded Los Angeles Free-Net, a nonprofit organization providing extensive medical information online. "Just a moment." Avrum interrupted my story. "I believe I have seen a paper documenting the movement of meningioma out of the brain. Let me see what I can find out for you." Within twenty-four hours I had a faxed medical paper at my office, documenting two such cases. How is it, I wondered, that my mom's oncologists and surgeons, among the very best in the world (and all remarkably humane throughout this ordeal), attending her at the USC/Norris Comprehensive Cancer Center and Hospital, one of the leading institutes for cancer research and treatment in the world, did not know about these cases? Worse, I thought, what if I had not gone to Carol's dinner party that night? Luck favors the hungry mind.

Lung surgery in October 1999, and a fifth craniotomy that also permanently removed an infected skull plate in January 2000, kept our emotional balance sheet in the black, but it was short-lived. The invasive malignant meningioma was on the march again.

In addition to the five craniotomies, my mom endured four gamma-knife radiosurgery treatments (September 1996, February 1998, January 1999, and March 2000), utilizing a powerful machine resembling a nineteenth-century phrenology contraption that fits tightly over the head, which can zero in on a pinpoint spot in the brain from all sides of the head, killing the cancer cells with a minimal amount of damage to the surrounding tissue. Finally, we were told that the risks of additional surgery and radiation outweighed the benefits. There was nothing more to be done.

What is a skeptic to do? An ideological commitment to science is one thing, but this was my mom! I turned to the literature on experimental drugs and, with the help of Avrum, determined that we would try mifepristone, a synthetic antiprogestin better known as RU-486 that operates by blocking the action of progesterone, a sex hormone that prepares the womb for the fertilized ovum. At the time, RU-486 was illegal, so we had to go through a Washington, D.C.–based women's rights advocacy group—the Feminist Majority Foundation's Mifepristone Compassionate Use Program—to obtain treatment dosages. A small-sample study suggested that it might retard the growth of tumors. It didn't for my mom, so we then tried temozolomide, a cytotoxic agent designed to prevent the replication of cells that divide rapidly, such as cancer cells. That proved equally ineffective.

The problem with such experimental treatments is that by the time patients get to the point of needing them their cancer is so far advanced that normal treatment dosages are most likely to be ineffectual. And, in any case, what works on genetically identical white mice may not work on genetically diverse humans. Finally, the sample sizes of experimental trials are typically small—one on temozolomide, for example, had thirty-one patients, one of which improved, two of which quit because of drug toxicity, and twenty-four of which "stopped treatment because of disease progression." That's hardly encouraging.

Nothing we tried was able to halt the inexorable march of the cancer that migrated into the middle of Mom's brain, where it could not be resected or radiated. The MRI diagnostic report was sterile but bone chilling: "Again identified is an approximately 2.5 x 4.0 cm mass in the right frontal lobe involving the corpus collosum. A similarly sized irregular mass is again identified in the left frontal lobe." My mom was dying. There was nothing to lose in trying some alternative cancer treatments, right?

The world of alternative and complementary medicine is a complex and murky one, particularly in the cancer community, involving countless unsubstantiated claims alongside glowing testimonials from cancer survivors (dead men tell no tales here either). As a former professional bicycle racer I could not be prouder of the fact that Lance Armstrong won the

Tour de France six times after coming back from death's door with testicular cancer that had trekked to his lungs and brain. It shall always reside in the annals of human achievement as akin to a miracle (although Lance attributes it entirely to good science, medicine, and training, as do I). But what about all those other cancer patients who never rode a bike again? We don't hear about them. This is an inherent problem in many fields of study: we remember the hits and forget the misses, celebrate the triumphs and bury the defeats, publish the successes and file-drawer the failures. (In scientific research, in fact, this is called the file-drawer problem, in which journals are inclined to publish only positive results. Negative results—the discovery, for example, that something does not cause cancer—usually get filed away as failed experiments.)

When faced with a grim prognosis we are typically offered a choice between scientific medicine that doesn't work and alternative medicine that might work. In fact, as I discovered in the very agonizing process of trying to save my mother's life, there is only scientific medicine that has been tested and everything else that has not been tested. Alternative medicine is not a matter of everything to gain and nothing to lose. There is much to lose, in fact, as I came to realize at a deep emotional level. Schlepping my dying mother around the country grew less appealing as I realized that the treatments might kill her before the cancer did. (Not to mention what a boatload of medications does to one's stamina and energy. Mom was downing a dozen different drugs every day, all of which played havoc with her biochemistry.) Given the ultimate fate of all flesh, what really would be accomplished in running down such chimera? My dad and I decided that it would be best if we spent as much quality time with Mom as we could.

The end came sooner than we thought. The tumors (or possibly the temozolomide) upset Mom's balance and she began falling—a risky thing when one is missing half a skullcap. We watched her as closely as was practical, but in the middle of the night she stumbled and fell headfirst into the corner of a television set, knocking herself unconscious. At four in the morning, waiting in an emergency room as a medical team tried to resuscitate her, it struck me as almost absurd after all of this to have the staff psychologist inquire if I would like to talk about my feelings. At least she didn't suggest an antidepressant.

Mom's body recovered, but her brain did not. She never regained full consciousness, although she could seemingly answer questions through hand squeezes. Someone was in there, so we kept her alive with a feeding tube. But after many agonizing weeks we arrived at the painful decision to pull the tube and start death's clock. It took a little over a week. The end came at 7:00 P.M. on Saturday, September 2, 2000. My dad, Richard, my wife, Kim, and my daughter, Devin, were present with me, holding Mom, telling her that it was okay to let go, and that we loved her. She was not conscious her final day so, as far as I know, she did not experience death. Instead, she slipped into unconsciousness, or some sort of altered state, until her body finally gave out. I believe it was a painless parting, and I am wholly comforted by the fact that just days before she was able to respond by squeezing our hands, so I took advantage of her awareness, however partial it might have been, by repeating over and over, dozens of times, "I love you, Mama. You were a great mother and a heroic person. Wise owl, brave soul. A life well lived. It's okay to say good-bye."

The surgeries, radiation treatments, and drug therapies were brutally hard on Mom, but as she told her sister, my beloved aunt Mary, in a touching handwritten note in a greeting card (the time and occasion for which I do not know), "We've shared lots of good times too! Remember—Tough times don't last, but tough people do." My mom was tough to the end. Through excruciating physical trauma she remained heroic and graceful, maintaining her perspective on life as a search for meaning through worthy challenges, authentic relationships, and the desire to be fair to all people and to always do the right thing. She did so not because it was required or expected, not for external rewards or public recognition, but because she would be a better person for it, and because we would all be better for it.

Too often, I think, we gloss over the messiness of living and the unpleasantries of life, particularly at the beginning and end, as if birth and death are shadowlands accessible only to a chosen few. We suppress or ignore some of the deepest and most meaningful events of the human condition—denial is not just a river in Egypt. It is in those shadowlands, however, where we face the termini of life and share the full experience of the hundred billion who came before us and know authentically what it means to be human.

Figure 7.1.   The author's mother as a young woman, the author and his mother at graduation, and his mother and father, just before the end.

PART III

SCIENCE AND THE (RE)WRITING OF HISTORY

8

Darwin on the *Bounty*

The How and the Why of the

Greatest Mutiny in History

> "You would have made an excellent historian;
> you have a profound contempt for facts."
>
> —the character of William Bligh as portrayed by
> Charles Laughton in the 1935 film *The Mutiny on the
> Bounty*, grilling an officer over missing coconuts

IN THE EARLY MORNING hours of April 28, 1789, one month after departing the South Pacific island of Tahiti, the crew of the HMS *Bounty*—a tiny merchant ship loaded with over a thousand pots of breadfruit trees bound for the Caribbean islands where they would be delivered as cheap slave fodder—awoke to the shouts and screams of a handful of men led by the ship's master's mate, Fletcher Christian, determined to overthrow the command of William Bligh and return the ship to Tahiti. Over the course of the next couple of hours the mutiny on the *Bounty* unfolded, resulting, over the course of the next couple of centuries, in an unforgettable tale that has taken on a life of its own and grown to mythic proportions.

Thousands of articles, hundreds of books, and five feature films have attempted, with varying degrees of success, to capture the gravitas of the mutiny on the *Bounty*, one of the most notorious events in naval history. Although the mutiny historiography is compendiously rich, I contend that the various theories proffered to explain the event and its participants all operate at a proximate historical causal level. Since the event has primarily been recounted and analyzed by historians who operate at this immediate level of who did what to whom and when, this makes sense. I suggest that an ultimate evolutionary causal level of analysis illuminates additional motives of Bligh, Christian, and the mutineers and offers a deeper understanding of why the mutiny really happened.

The Myth of the Mutiny

The most common explanation for the mutiny on the *Bounty*, which arose about a decade after the event and is still the one most often cited, pits an oppressive William Bligh against a humane Fletcher Christian. In the name of justice, the myth holds, Christian reluctantly rebelled against Bligh's totalitarian regime, casting him adrift after countless tirades and undeserved floggings, and returning his tortured men to the freedom offered in the South Pacific. Bligh survived the voyage home, while Christian and the mutineers lived out their lives in peace on Pitcairn Island, thousands of miles from nowhere.

The myth busting begins at the top with the tyrannical William Bligh who, as it turns out under closer inspection, wasn't. The Australian historian Greg Dening, in his 1992 narrative history, *Mr. Bligh's Bad Language*, undertook a complete count of every lash British sailors received on fifteen naval vessels that sailed into the Pacific from 1765 to 1793. Of the 1,556 sailors, 21.5 percent were flogged. The celebrated Captain James Cook, for example, flogged 20, 26, and 37 percent respectively on his three voyages; the distinguished explorer Captain George Vancouver (of Vancouver Island fame) flogged 45 percent of his charges; Bligh comes in at 19 percent on the *Bounty*, and only 8 percent of his subsequent *Providence* voyage. Dening computes a mean of 5 lashes for every sailor in the cohort. By comparison, Vancouver's mean was 21 lashes, Bligh's *Bounty* mean a measly 1.5 lashes.

If draconian punishment was not the cause of the mutiny, then what was? The most recent revisionist explanation is Caroline Alexander's 2003 book, *The Bounty* (subtitled, in pre-postmodern style, "The True Story of the Mutiny on the *Bounty*"). Alexander's book is a comprehensive narrative that utilizes such primary documents as the court-martial trial transcripts and some of the mutineers' private diaries and letters. In the

Figure 8.1.   William Bligh

Figure 8.2.   Fletcher Christian

process, she identifies the source of the tyranny myth with the families of two of the mutineers, who spin-doctored Bligh into the foaming-at-the-mouth character so portrayed by Charles Laughton in the 1935 film interpretation. In Alexander's account, Fletcher Christian was the *Bounty* antihero, not Bligh, who emerges in her telling as one of the greatest seafaring commanders in naval history. Yet, after four hundred pages of gripping narrative, Alexander wonders "what caused the mutiny," hints that it might have had something to do with "the seductions of Tahiti" and

"Bligh's harsh tongue," but then concludes that it was "a night of drinking and a proud man's pride, a low moment on one gray dawn, a momentary and fatal slip in a gentleman's code of discipline." That is the extent of her causal analysis.

Cowards and Causes

Dening also blanches when it comes to the search for ultimate causes, even retreating from the quantitative science he employed so effectively in debunking the harsh-Bligh myth. In its stead he turns to literary analysis. According to Dening, the numerous retellings of the mutiny on the *Bounty* in literature and film say more about the cultures of the authors and the filmmakers than they do about the actual mutiny. Charles Laughton's Bligh and Clark Gable's Christian in the 1935 film presented a tale of class conflict and of tyranny versus justice, allegedly reflecting 1930s America. The 1962 film, with Trevor Howard as Bligh and Marlon Brando as Christian, presented the mutinous conflict as one of naked profit-seeking versus humane and liberal values, supposedly mirroring the changing morals in the transition from the conservative 1950s to the liberated 1960s. The 1984 film purportedly flirts with a homosexual theme, with Anthony Hopkins's Bligh at once attracted to but outraged by Mel Gibson's Christian, who does not return his affections.

To one who has seen all the films and read many of the books, this is all very interesting, but *what actually happened* on the *Bounty*? Here Dening can take us only so far because, he says, at bottom history is nothing more than an echo of the historian's times, an "illusion," he calls it, of a past that can only ever be a reflection of our present. Dening, in fact, tells his students: "History is something we make rather than something we learn. . . . I want to persuade them that any history they make will be fiction." But the mutiny was no fiction, and one cannot write a nonfiction work without offering some causal thesis, of which Dening's is Bligh's bad language. *Well shiver me timbers and blow me down mates!* Imagine that—a sailor using bad language.

By language, however, Dening means something more than obscenities. He means Bligh's use of language as a means of communicating his intentions, a shortcoming of which so disrupted the hierarchical relationships on the ship that it led to mutiny. (An editorial cartoon lampooning this thesis shows Bligh drifting away in the *Bounty*'s launch, shouting, "So, Mr. Christian! You propose to unceremoniously cast me adrift?" The caption reads: "The crew can no longer tolerate Captain Bligh's ruthless splitting of infinitives.") Bligh, Dening says, "found it difficult to grasp the metaphors of being a captain, how it could mean something different to those being captained," and "tended not to hear the good intentions or catch the circumstances and context in the language of others but demanded that others hear them in his."

Dening is on to something here, because his thesis taps into the psychology of status and hierarchy, which we shall consider below. But it strikes me as supremely ironic to note that if we were to employ Dening's historiographical methods to his own work we would have to note that he very much reflects his own postmodern, deconstructionist culture of the 1980s and 1990s, in which textual analysis and theories of language specify how we should "read" history. This is precisely what Dening does in laying the blame on Bligh's language. Of course, Dening says history is nothing more than a reflection of the historian's culture, so we should not be surprised that he has taken this approach. But here Dening has a problem. In order to convince the reader that his methodology is superior to others', he must reject the earlier theories about the mutiny and prove that his is correct. To do so he presents objective evidence, such as the number of floggings. In other words, Dening must temporarily abandon his own theory of history in order to support it. So much for postmodern historiography, which leads to such positions as Dening's, who confesses: "Debate on why there was a mutiny on the *Bounty* has been long. Who can—who would want to— end it? Not I. I am a coward for causes."

Fortunately for history, scientists are champions for causes, of which there are two types: (1) *proximate* (historical) and (2) *ultimate* (evolutionary).

Proximate Causes

The *Bounty* was originally called the *Bethia*, a 215-ton merchant ship, 90 feet 10 inches long, 24 feet 3 inches wide, with a total crew of only fifteen men. For Bligh's mission the little ship was refitted and renamed, with the great cabin and other spaces converted into storage spaces for the breadfruit plants. To procure, maintain, and deliver the cargo, the crew was expanded threefold to forty-five, all volunteers (unusual for the British navy of the time, for which most sailors were pressed into service). With even less crew space than the *Bethia* provided, the result was extreme overcrowding, adding to the tensions already building from other causes. Despite the crew expansion, the ship lacked the customary marine guards to protect the captain, and although Bligh was the *Bounty*'s commander, he was not promoted to captain because the British admiralty determined that the ship's diminutive size did not warrant it, a decision that would cause Bligh to push himself and his charges harder than he might have had his confidence in his authority on board, and hierarchy within the navy, been secured.

The *Bounty* departed Portsmouth on December 23, 1787. Bligh wanted to sail to Tahiti by way of Cape Horn at the tip of South America, in order to complete a circumnavigation of the globe, a feat he hoped would enhance his climb up the naval chain of command. But because of over a month in delays in getting under way, the *Bounty* arrived at the Cape during the high storm season. After a month of being pounded by high seas and thunderous rains, Bligh relented and turned the ship eastward toward Africa and the Cape of Good Hope, where they would find their way to Tahiti in a different (but less glorious) direction.

The *Bounty* finally arrived in Matavai Bay, Tahiti, October 26, 1788, for a total sailing time of ten months, three days, and a remarkable 27,086 miles. Five months later, the ship departed Tahiti, on March 28, 1789, a longer stay than planned due to the failure of a sufficient number of breadfruit seedlings to sprout. During this delay, many of the men participated fully in village life, including taking on native lovers. Fletcher Christian, who was put in charge of collecting the breadfruit plants, was reportedly among those most eager to embrace local customs, including procuring for himself a number of Polynesian tattoos (including those traditionally required for marriage eligibility) and taking on a permanent lover with whom he cohabitated.

On the return voyage tensions mounted. Bligh and Christian led a shore party to the small island of Anamooka to procure fresh water, but Christian's group had been thwarted by the natives and Bligh damned him "for a Cowardly rascal, asking him if he was afraid of a set of Naked Savages while He had arms," crewman James Morrison later recalled, "to which Mr. Christian answered 'the Arms are no use while your orders prevent them from being used.'" (Bligh had instructed the men not to bear arms directly because it might incite the natives, but instead keep them nearby just in case. The result was that the natives surrounded the men, stole some tools, and chased them away from the watering hole.) Christian grew ever more restless, confiding to midshipman Edward Young that he wanted to build a raft and paddle to a small island the ship was passing, then make his way back to Tahiti. Young pointed out that the sharks would probably nab him before Bligh could, and Bligh later penned his opinion of the plan: "That Christian . . . intended to go onshore 10 leagues from the land on a fair Plank with two staves for Paddles with a roasted Pig is too ridiculous." Christian hatched an alternative plan and bided his time.

On the calm clear evening of April 27, by the glow of a volcanic eruption from the island of Tofua in the distance, Bligh came on deck and noticed that some coconuts had gone missing. He immediately called for the master of the ship, John Fryer, who recalled that Bligh admonished, "Mr. Fryer, don't you think that those Cocoanuts are shrunk since last Night?" Fryer thought not, but Bligh insisted "that they had been taken away and that he would find out who had taken them." In a scene reminiscent of that from the film *The Caine Mutiny*, in which Humphrey Bogart's Captain Queeg has a paroxysm over missing strawberries and turns the ship upside down looking for them, Bligh called all officers on deck and grilled them one by one. According to James Morrison, Bligh "questioned every Officer in turn concerning the Number they had bought, & coming to Mr. Christian askd Him, Mr. Christian answerd 'I do not know Sir, but I hope you don't think me so mean as to be Guilty of Stealing Yours.' Mr. Bligh replied 'Yes you dam'd Hound I do—You must have stolen them from me or you could give a better account of them—God dam you you Scounderels you are all thieves alike, and combine with the men to rob me—I suppose you'll Steal my Yams next, but I'll sweat you for it you rascals I'll make half of you jump overboard before you get through Endeavour Streights.'" Oddly, in an apparent gesture of reconciliation, Bligh granted the men a pound and a half of yams each, but then threatened to reduce it to three-quarters of a pound if the coconuts were not returned.

Emotions were riding high. William Purcell later recalled that Christian came away from the confrontation with tears "running fast from his eyes in big drops." He inquired, "What is the matter Mr. Christian?" Christian replied: "Can you ask me, and hear the treatment I receive?" Yet

Figure 8.3.   The mutiny on the *Bounty*

Bligh again demurred, inviting Christian to dine with him that evening. The invitation was declined. Christian had something else in mind for that night.

All was quiet during the first two watches of the night, from eight to midnight, and from midnight to 4 A.M., when Fletcher Christian took the watch. An hour later, around 5 A.M., as the first light of the sun warmed the horizon, Christian and several others broke into the arms' chest to procure some weapons, then moved to Bligh's cabin, awoke him, tied his hands behind his back, and took him up on deck. Men on both sides were screaming and shouting, threatening and gesturing. Sometime amid the chaos, Christian was heard to tell Bligh: "Sir your abuse is so bad that I cannot do my Duty with any Pleasure. I have been in hell for weeks with you." Details of the mutiny varied from eyewitness to eyewitness, but all remembered Christian's haunting description of his torment. Bligh implored him, "Consider Mr. Christian, I have a wife and four children in England, and you have danced my children upon your knee." Christian's rejoinder was unequivocal: "That!—captain Bligh,—that is the thing—I am in hell—I am in hell."

Christian then had Bligh and eighteen men placed in the *Bounty*'s twenty-three-foot launch, along with whatever possessions and supplies they could nab on the way to what appeared to be their doom: 150 pounds of bread, 32 pounds of pork, 6 quarts of rum, 6 bottles of wine, and 28 gallons of water. It was enough to support them for five days. They lasted forty-eight, covering a remarkable 3,618 miles without a single loss of life, in what has come to be regarded as the single greatest feat in ocean sailing. As the two ships diverged, the mutineers were heard to shout, "Huzza for Otaheiti." Their worlds were forever rent asunder.

Although the immediate cause of the mutiny was the great coconut brouhaha, this is surely no cause for the ultimate high-seas crime. Why would an experienced commander like Bligh overreact to such a minor incident? Why would a hardy sailor like Christian take Bligh's reproach so personally that it drew tears? Why was Fletcher Christian "in hell" and what role did Bligh play in putting him there?

Ultimate Causes

The emotions expressed in the mutiny were more than those triggered by immediate circumstances; there was a deeper foundation to them, of which we shall consider two: (1) bonding and attachment and (2) hierarchy and status.

Bonding and Attachment

With countless hours to reflect on what happened and why (it took him 10.5 months to wend his way back to England), Bligh worked through the proximate causal explanations of who did what to whom and when, and began to search for ultimate causes. Why would his men revolt against him? In his *Narrative of the Mutiny on Board His Majesty's Ship Bounty*, which he penned during the open-boat voyage home, Bligh began to understand a deeper cause of the mutiny.

> It is certainly true that no effect could take place without a Cause, but here it is equally certain that no cause could justify such an effect— It however may very naturally be asked what could be the

reason for such a revolt, in answer to which I can only conjecture that they have Idealy assured themselves of a more happy life among the Otaheitians than they could possibly have in England, which joined to some Female connections has most likely been the leading cause of the whole business.

Although predating Darwin by a century, Bligh understood that the youth of his men meant that they had few permanent and lasting commitments at home, and that they could easily form bonds and attachments with the women in Tahiti.

The Women are handsome—mild in their Manners and conversation—possessed of great sensibility, and have sufficient delicacy to make them admired and beloved—The chiefs have taken such a liking to our People that they have rather encouraged their stay among them than otherwise, and even made promises of large possessions. Under these and many other attendant circumstances equally desirable it is therefore now not to be Wondered at, 'tho not possible to be foreseen, that a Set of Sailors led by Officers and void of connections, or if they have any, not possessed of natural feelings sufficient to wish themselves never to be separated from them, should be governed by such powerfull inducement but equal to this, what a temptation it is to such wretches when they find it in their power however illegally it can be got at, to fix themselves in the most of plenty in the finest Island in the World where they need not labour, and where the allurements of dissipation are more than equal to anything that can be conceived.

Consider the makeup of naval crews in this period. They consisted of young men in the prime of sexual life, designed by evolution to pair-bond in serial monogamy with women of reproductive age. Of the crews who sailed into the Pacific from 1765 to 1793, 82.1 percent were between the ages of twelve and thirty, and another 14.3 percent between the ages of thirty and forty. The average age of the *Bounty* crew was twenty-six. Bligh was thirty-three. There were, in fact, seven men older than Bligh (all thirty-six or more; one was thirty-nine), making the *Bounty*'s crew the oldest to sail into the Pacific during this period. When they arrived at the South Pacific they exhibited little self-restraint. Of the 1,556 sailors in this general cohort, 437 (28 percent) got the "venereals." The *Bounty*'s crew was among the highest at 39 percent (with Cook's *Resolution* and Vancouver's *Chatham* crews topping out at 57 and 59 percent, respectively). Female connections were rampant. Fletcher Christian was one of those treated for "venereals."

After the mutiny Christian sailed the *Bounty* back to Tahiti, and on September 23, 1789, he left the island, taking with him eight mutineers, six Tahitian men, eleven Tahitian women, and one child. On January 15, 1790, they arrived at Pitcairn Island, unloaded the ship, and a week later torched it in a gesture of no return. On a Thursday that October, to Fletcher Christian and his Polynesian wife was born a son, which he named Thursday October Christian. Each mutineer, in fact, had taken a woman as a mate, but only three of the six Polynesian men were attached. The consequences of this inequity were predictable. After three years, the woman living with mutineer John Williams died, so he took a replacement from one of the Polynesian men, leading to jealousy, violence, and retribution, and a massacre on September 20, 1793, in which five of the mutineers were killed, including Fletcher Christian, as well as all of the Polynesian men. In subsequent years, William McCoy committed suicide, Matthew Quintal was killed by one of the other mutineers, and Edward Young died of asthma. By 1800, the only male survivor was John Adams, who lived until 1829, long enough to tell the tale of the fate of the *Bounty*'s mutineers. One such account, retold by Captain Philip Pipon on the HMS *Tagus*, offered this corroboration of Bligh's insight into Christian's motives:

> Christian's wife had paid the debt of nature, & as we have every reason to suppose sensuality & a passion for the females of Otaheite chiefly instigated him to the rash step he had taken, so it is readily to be believed he would have lived long on the island without a female companion.

Another British captain who visited Pitcairn in 1825, Frederick Beechey, interviewed Adams, who recalled that Christian relieved the officer on duty and, turning to Quintal, "the only one of the seamen, who Adams said, had formed any serious [female] attachment at Otaheite," cajoled him to consider that "success would restore them all to the happy island, and the connexions they had left behind." As Caroline Alexander describes it, it was "not just a life of ease" they were leaving behind, "but friends, lovers, common-law wives, in some cases their future children."

There are certain basic facts about human nature that emerge with even casual observation: we are a hierarchical social primate species, relatively avaricious, self-serving, and desirous of possessions, and one of the most sexual of all primates, where males are more openly and intensely obsessed with obtaining sexual unions and equally preoccupied with protecting those unions from violation by other males. It doesn't take Darwin, or even modern evolutionary psychology, to be cognizant of these facts. Organized religion figured them out millennia ago, and they emerge in such lists of immoral behaviors as the seven deadly sins: pride, envy, gluttony, lust, anger, greed, and sloth; followed by their moral antonyms: humility, kindness, abstinence, chastity, patience, liberality, and diligence (and, for good measure, the theological virtues of faith, hope, and charity, along with the cardinal virtues of fortitude, justice, temperance, and prudence). This is one reason for the need for strict discipline on board overcrowded ships, regulated by hierarchy and status.

The evolutionary foundation for the mutineer's heightened state of emotions is borne out in modern neuroscience, which shows that the attachment bonds between men and women are powerful forms of chemical addiction, especially in the early stages of a relationship. Dopamine is a neurotransmitter substance secreted in the brain that, according to Helen Fisher in her book *Why We Love,* produces "extremely focused attention, as well as unwavering motivation and goal-directed behaviors." Dopamine is elevated in the brains of animals intensely bonded to another individual, and in humans "elevated concentrations of dopamine in the brain produce exhilaration, as well as many of the other feelings that lovers report—including increased energy, hyperactivity, sleeplessness, loss of appetite, trembling, a pounding heart, accelerated breathing, and sometimes mania, anxiety, or fear." Attachments this strong are literally addictive, stimulating the same region of the brain that is active in drug addictions. A related brain chemical, norepinephrine, is also associated with bonding and attachment, and it too produces "exhilaration, excessive energy, sleeplessness, and loss of appetite." Another hormone, oxytocin, is also associated with attachment. In her book *The Oxytocin Factor*, Kerstin Uvnäs Moberg shows that oxytocin is a hormone secreted into the blood by the pituitary during sex, particularly orgasm, and plays a role in pair-

Figure 8.4.
The women of Otaheiti

bonding, an evolutionary adaptation for long-term care of helpless infants. In women it stimulates contractions at birth, lactation, and mental bonding with the infant. In both women and men it increases during sex and surges at orgasm, possibly playing a role in pair-bonding—monogamous species secrete more oxytocin during sex than polygamous species.

Fisher also conducted brain scans (functional magnetic resonance image, or fMRI) of subjects in love, discovering that a structure buried deep in the brain (meaning it is evolutionarily ancient) called the caudate nucleus becomes extremely active when subjects gaze at the face of their lovers. Interestingly, the caudate nucleus is part of the brain's reward system. "The caudate helps us detect and perceive a reward, discriminate between rewards, prefer a particular reward, anticipate a reward, and expect a reward. It produces motivation to acquire a reward and plans specific movements to obtain a reward." And, most interestingly, Fisher's fMRI experiments found heightened activity in the ventral tegmental area of the brain, which is a center for dopamine-making cells, the same dopamine that surges during heightened attachment.

Now, it is too simple to conclude that the mutiny on the *Bounty* happened because Christian and his followers were desperate to return to their new attachments in Tahiti. In the context of spending ten months at sea, which attenuated already weak home attachments, and of forming new and powerful bonds made through sexual liaisons in Tahiti (that in some cases even led to cohabitation and pregnancy), the men experienced elevated emotions that culminated in a breaking point of revolt one month after departing the island, just as the men reached peak withdrawal from their attachment addictions. Any excuse might suffice, and Bligh's tirade over the stolen coconuts was one such trigger. For his part, Bligh was happily married and showed no interest in the native women, but he had a different force tugging at his brain. Bligh's already hair-trigger temperament was put on alert by his drive to preserve hierarchy on board and to gain status in his career.

Hierarchy and Status

We have already seen that Bligh was frustrated from the start of the voyage by not being promoted to captain. Some background on Bligh reveals what effect this might have had on him, and some additional material on our scientific understanding of the role of hierarchy and status fills in our ultimate understanding of the mutiny.

Born on September 9, 1754, the firstborn son of a custom's officer, William Bligh developed into a short, stocky man with black hair, blue eyes, and a reputation for using foul language when angered. He first went to sea at age sixteen as an able-bodied seaman, was promoted to midshipman after seven months, then worked his way up the officer ranks.

At age twenty-three Bligh sailed with the renowned Captain James Cook as a commissioned warrant officer and navigator aboard the HMS *Resolution*, Cook's third voyage (during which he was killed in the Sandwiche Isles in 1779, on what is today the big island of Hawaii). Bligh skillfully and heroically navigated the *Resolution* back to England but did not receive the credit he thought he deserved for this, or for his general mapping and surveying tasks during the voyage. And had Cook survived, Bligh likely would have been granted his long-sought captainship, since such promotions were typically pushed through by a mentor; without Cook, Bligh was denied promotion.

During the war with France, Bligh served with distinction and was promoted to lieutenant in 1781. He was also married that year to the daughter of a custom's officer and appointed as master of HMS *Cambridge*, on which he first met Fletcher Christian. In his diary, Christian claimed that Bligh "treated him like a brother." Bligh taught Christian how to use a sextant for navigation, and they frequently dined together. When peace broke out in 1783, the navy's budget was cut and Bligh was reduced to half pay. He took command of a merchant ship named *Britannia* and sailed between England and the West Indies, during which he once again commanded Fletcher Christian.

When he took command of the *Bounty* as her only commissioned officer, Bligh took a severe cut in pay—from 500 pounds a year to only 50 pounds a year—in hopes that a successful voyage would pay off in a higher rank. Bligh's goal was to circumnavigate the globe, deliver the breadfruit, and return with no punishments delivered and no lives lost. By most accounts he was a professional, humanitarian leader. For example, concerned about the dreaded seaman's disease scurvy, he regularly served the men sauerkraut. He understood the importance of physical activity beyond work, so he arranged for a near-blind fiddler named Michael Byrne to play music to which the men could dance on deck. To attenuate the drudgery of work, he split the crew into three watches instead of the usual two, which gave them more breaks.

On the way to Tahiti, Bligh administered only one flogging, about which he wrote: "Until this afternoon I had hoped I could have performed this voyage without punishment to anyone." The flogging was of Matthew Quintal, charged by the master of the ship, John Fryer, with "mutinous behaviour," which Bligh munificently downgraded to "insolence and contempt." Because of the excessive time spent getting to Tahiti, while on the island Bligh was lenient with his charges. When he delegated orders, he did not always follow up to confirm that they were executed. The ship was not taken out for routine tune-up trips, an anchor line broke, and the sails began to rot in the heat and humidity. On January 5, 1789, three crew members deserted—William Muspratt, John Millward, and Charles Churchill—stealing the ship's launch along with muskets. Thomas Hayward was officer of the watch, who claimed he was asleep when it happened. Bligh had him confined in irons, then set out to capture the deserters, which took him three weeks. Churchill received a dozen lashes; the other two got two dozen each; pace the draconian Bligh myth, all were let off easy since the recommended punishment for desertion was death by hanging.

As for Bligh's infamous temper—he called his officers "scoundrels, damned rascals, hounds, hell-hounds, beasts and infamous wretches"—one of his defenders, George Tobin, who sailed with Bligh on the *Providence* in 1790, observed: "Those violent tornados of temper when he lost himself" were upsetting, and "Once or twice, indeed, I felt the unbridled license of his power of speech, yet never without soon receiving something like a plaister to heal the wound . . . when all, in his opinion, was right, who could be a man more placid and interesting."

Bligh was walking a narrow plank between enforcing the strict rules of the British navy—refined over the centuries for maintaining discipline among young men stuck in tight quarters for long periods at sea—and earning the respect of his men through a fair and humanitarian command. His temper was a by-product of the struggle to find this balance, along with fulfilling his own personal ambition. In 1805, in another court-martial trial for "unofficerlike conduct and ungentlemanly behavior," Bligh admitted that his "high sense of professional duty" made him "sometimes too particular in the execution of it." In fact, Bligh's harsh temperament was directed not to the lower-ranking men but at his commissioned, warrant, and petty officers—those responsible for the rank and file, and from which his hierarchy was dependent down the chain of command. For example, Bligh snapped at a lieutenant: "What, Sir, you damn'd scoundrel, never was a man troubled with such a lot of blackguards as I am. Take care, Sir, I am looking out for you." To the court-martial judges he explained the context and motives for such comments:

> I candidly and without reserve avow that I am not a tame and indifferent observer of the manner in which officers placed under my orders conduct themselves in the performance of their several duties. A signal or any communication from a commanding officer has ever been to me an indication for exertion and alacrity to carry into effect the purport thereof and peradventure I may occasionally have appeared to some of these officers as unnecessarily anxious for its execution by exhibiting an action or gesture peculiar to myself to such.

Let's not forget the status and hierarchy of the other major player in this drama, Fletcher Christian. Born in Cumberland on September 15, 1764, Christian was ten years younger than Bligh, a last born whose older brothers, John and Edward, were well educated at Peterhouse and St. John's College, Cambridge. Edward became a professor of law at Cambridge (Bligh called him "a sixpenny Professor"). Like Bligh before him, Fletcher first went to sea at age sixteen, and two years later he sailed under Bligh on the HMS *Cambridge*. At five feet nine inches, he was tall, and was sometimes described as "swashbuckling," "a slack disciplinarian," and "a ladies man" who was conceited but mild, generous, and open. In recounting the mutineers in his narrative, Bligh described Christian as "master's mate, aged twenty-four years, five feet nine inches high, blackish, or very dark brown complexion, dark brown hair, strong made, a star tattooed on his left breast, tattooed on his backside; his knees stand a little out, and he may be called rather bow-legged. He is subject to violent perspirations, and particularly in his hands, so that he soils anything he handles."

Fletcher applied to be master of the *Bounty*, but John Fryer was already appointed by the admiralty, so he mustered as master's mate, along with William Elphinstone, so at most Christian coshared third in command. During the voyage, however, Bligh promoted him to acting lieutenant, which made him second in command over Fryer. So Bligh wasn't the only officer on board with social status ambitions. For Christian, a successful *Bounty* mission might net him even more than a couple of higher notches up the naval hierarchy; within his family niche he stood to gain status among his already successful older brothers.

In this Bligh and Fletcher were following their evolutionary impulse to gain status. The overwhelming evidence from anthropologists is that most social mammals, all social primates, and every human community ever studied show some form of hierarchy and social status. While it is true that hunter-gatherer communities are much more egalitarian than modern state societies, sans such a dramatic comparison one can find subtle but real forms of social hierarchy in all peoples. The African pygmies called Aka, for example, are relatively egalitarian, but they recognize a leader, called a *kombeti*, who is typically a highly skilled hunter with prestige and power within the group, who is rewarded with more food, women, and children. In the South American Ache peoples, meat is distributed and consumed equally among all members of the community, but the most successful hunters have more extramarital affairs and more children— legitimate and illegitimate—than less successful hunters; and those children survived better than the offspring of other hunters. Anthropologist Donald Brown even includes social hierarchy on his list of human universals, meaning there are no exceptions.

Note, as well, the association of bonding and attachment with hierarchy and status. The higher your status and hierarchy, the greater your opportunity for bonding and attachment. As Darwin observed in *The Descent of Man*:

> The strongest and most vigorous men,—those who could best defend and hunt for their families, and during later times the chiefs or head-men,—those who were provided with the best weapons and who possessed the most property, such as a large number of dogs or other animals, would have succeeded in rearing a greater average number of offspring, than would the weaker, poorer and lower members of the same tribes. There can, also, be no doubt that such men would generally have been able to select the more attractive women. At present the chiefs of nearly every tribe throughout the world succeed in obtaining more than one wife.

Anthropologist Napoleon Chagnon documented just such an effect among the Yanomamö people of Amazonia, in which the most successful warriors fathered the most number of offspring.

> If you take men who are in the same age category and divide them by those who have killed other men (unokais) and those who have not killed other men (non-unokais), in every age category unokais had more offspring. In fact, unokais averaged 4.91 children versus 1.59 for non-unokais. The reason is clear in the data on the number of wives: unokais averaged 1.63 wives versus 0.63 for non-unokais. This was an unacceptable finding for those who hold the ideal view of the Noble savage. "Here's Chagnon saying that war has something good in it." I never said any such thing. I merely pointed out that in the Yanomamö society, just like in our own and other societies, people who are successful and good warriors, who defend the folks back home, are showered with praise and rewards. In our own

culture, for example, draft dodgers are considered a shame. Being a successful warrior has social rewards in all cultures. The Yanomamö warriors do not get medals and media. They get more wives.

It's not that one is consciously seeking status in order to attract more mates and intimidate competing males; the effect is a subtle one, involving a complex array of thoughts and emotions. Thus, we shouldn't be surprised to learn that there is a biochemical connection here as well. Recall the brain chemicals that turn love into an addiction. In brain studies a connection has been found between seeking social status and the neurotransmitter serotonin. In African vervet monkey groups, the highest-ranking males have the highest level of serotonin. In humans, serotonin makes people more sociable and assertive, extroverted, and outgoing. Low levels of serotonin are associated with low self-esteem and depression, and are countered pharmacologically with antidepressant drugs such as the SSRIs—selective serotonin reuptake inhibitors (e.g., Prozac)—which block the reuptake of serotonin in the synaptic gaps between neurons, causing them to fire more and make you feel more awake, alert, and alive!

Now, again, it would be too simple to conclude that Bligh's and Christian's competing desires for status led to the mutiny on the *Bounty*; but it is fair to consider that the dual effects of our two evolutionary forces—strong attachments in Tahiti, coupled with Bligh's and Christian's competition for status and hierarchy—together produced an underlying level of emotion triggered by the proximate events of April 27–28, 1789.

Proximate causes of the mutiny on the *Bounty* may have been missing coconuts and lost tempers, but the ultimate cause was evolutionarily adaptive emotions expressed nonadaptively in the wrong place at the wrong time, with irreversible consequences.

9

Exorcising Laplace's Demon

Clio, Chaos, and Complexity

HISTORIANS DO NOT COMMONLY turn to engineers for insight into how and why the past unfolds as it does. But since the process of modeling history in the language of chaos theory and nonlinear dynamics is a richly interdisciplinary study far afield from traditional historiography, it may be fruitful to seek the thoughts of one whose vision has been sharper than most.[1] Paul MacCready, inventor of human- and solar-powered flight, pioneer of electric automobiles, and engineer extraordinaire, upon reading a paper I had written on chaos and history,[2] found himself "thinking of a note I contributed to *American Scientist* wherein some 75 scientists described what got them into science. I was the only respondent to note that a person answering this question creates a plausible history, but not necessarily a real one."[3]

> There were innumerable influences in your past, but you remember only a few of the major ones, and you instinctively weave these into a plausible history explaining how you became what you presume you are. This interpretation of history is both logical and nonfalsifiable and so tends to establish its own validity. Chances are it's wrong.[4]

This problem of reconstructing the past to explain the present is, of course, an old and familiar one to philosophers of history but in this context needs restating because there is considerable risk in weaving plausible histories with instruments from another science. Are we, as William H. McNeill told me he once did, merely remodeling history to fit the language of a new physics? "I find your argument about chaos and history entirely congenial, though when I was young and struggled with finding my own ways of thinking about these questions I was drawn to the language of an older physics, seeing human society as a species of equilibrium, nested within other equilibria—biological, physical—and each level of equilibrium interacting with the others."[5] Is this no more than an epistemological game of chasing the latest trends in the physical sciences? If history changes with the physics of the day, then are we no closer to understanding what really happened and why? Perhaps, but then all of the sciences are equally guilty. Science is not the affirmation of a set of beliefs but a process of inquiry aimed at building a testable body of knowledge open to rejection or confirmation. So too is history . . . or at least it should be. Thus, making such applications from another field is no more and no less than what others have always done.

The integration of chaos and history is one of using a theory of present change to explain past change. Other historical scientists employ this strategy. In paleontology, Niles Eldredge and Stephen Jay Gould developed their evolutionary theory of punctuated equilibrium by applying a modern theory of speciation to the fossil record. (According to the theory of allopatric speciation, a new species is created when a small group of organisms is isolated from the larger population, then changes dramatically—larger populations are more genetically stable than smaller populations. The period of change is relatively rapid compared to the stability of long-lasting larger populations. In the paleontological record this process would leave few transitional fossils, leaving so-called gaps described by Eldredge and Gould as evidence of a speciational process.)[6] In archaeology, Lewis Binford argues that it is the job of the archaeologist not just to record the raw data of a dig but to interpret human action from human artifacts by the development of "principles determining the nature of archaeological remains to propositions regarding processes and events of the past."[7] This is what Fernand Braudel insists be done in the application of psychological, sociological, and economic theories to history.[8] Human history is human behavior writ past.

Chasing the Universal, Embracing the Particular

The issue of randomness and predictability in physical, biological, or social systems remains one of great import because the debate touches on such deeply meaningful issues as free will and determinism. Carl Hempel's search for so-called covering laws in history was an attempt to extract predictability through historical laws. Hempel went so far as to conclude that "there is no difference between history and the natural sciences."[9] Hempel was wrong about covering laws, but right about history and the natural sciences—not, however, in the direction one might think. History is not governed by Hempel's "universal conditional forms," but neither are the physical and biological worlds to the extent we have been led to believe. Scientists are coming to realize that the Newtonian clockwork universe is filled with contingencies, catastrophes, and chaos, making precise predictions of all but the simplest physical systems virtually impossible. We could predict precisely when Comet Shoemaker-Levy 9 would hit Jupiter but could muster at best only a wild guess as to the effects of the impacts on the Jovian world. The guess was completely wrong. Why? The answer strikes at the heart of understanding the nature of causality, as the late Stephen Jay Gould notes: "Do large effects arise as simple extensions of small changes produced by the ordinary, deterministic causes that we can study every day, or do occasional catastrophes introduce strong elements of capriciousness and unpredictability to the pathways of planetary history?"[10]

There is some irony in Gould's inquiry. For decades historians chased scientists in quest of universal laws but gave up and returned to narratives filled with capricious, contingent, and unpredictable elements that make up the past, resigning ourselves to the fact that we would never be as good. Meanwhile, a handful of scientists, instead of chasing the elusive universal form, began to write the equivalent of scientific narratives of systems' histories, integrating historical contingencies with nature's necessities. Ironically, says Gould, "This essential tension between the influence of individuals and the power of predictable forces has been well appreciated by historians, but remains foreign to the thoughts and procedures of most scientists."[11] Gould demonstrates how even a subject as predictable and subservient to natural law as planets and their moons, when examined closely, reveal so much uniqueness and individuality that while "we anticipated greater regularity . . . the surfaces of planets and moons cannot be predicted from a few general rules. To understand planetary surfaces, we must learn the particular history of each body as an individual object—the story of its collisions and catastrophes, more than its steady accumulations; in other words, its unpredictable single jolts more than its daily operations under nature's laws."[12] History matters. Narrative lives.

A Chaotic Model of Historical Sequences

Historians have been cognizant for millennia of the basic principles of chaos and nonlinearity. The principles, though, are coded in a different language and grouped into two fundamental "forces" that guide historical sequences—*contingency* and *necessity*. In this analysis contingency will be taken to mean *a conjuncture of events occurring without perceptible design,* and necessity to be *constraining circumstances compelling a certain course of action.* Contingencies are the sometimes small, apparently insignificant, and usually unexpected events of life—the kingdom hangs in the balance awaiting the horseshoe nail. Necessities are the large and powerful laws of nature and trends of history—once the kingdom has collapsed 100,000 horseshoe nails will not help a bit. Leaving either contingency or necessity out of the historical formula, however, is to ignore an important component in the development of historical sequences. The past is constructed by a dynamic interaction of both, and therefore it might be useful to combine the two into one term that expresses this interrelationship—*contingent-necessity*—taken to mean *a conjuncture of events compelling a certain course of action by constraining prior conditions.*

Randomness and predictability—contingency and necessity—long seen to be opposites on a quantitative continuum, are not mutually exclusive models of nature from which we must choose. Rather, they are qualitative characteristics that vary in the amount of their respective influence and at what time their influence is greatest in the chronological sequence. No one denies that such historical necessities as economic systems, demographic trends, geographical locals, scientific paradigms, and ideological worldviews exert a governing force upon individuals falling within their purview. Contingencies, however, exercise power sometimes in spite of the necessities influencing them. At the same time they reshape new and future necessities. There is a rich matrix of interactions between early pervasive contingencies and later local necessities, varying over time, in what is here called the *model of contingent-necessity*, which states: *In the development of any historical sequence the role of contingencies in the construction of necessities is accentuated in the early stages and attenuated in the later.*

There are corollaries that encompass five aspects of the model, including:

Corollary 1: The earlier in the development of any historical sequence, the more chaotic the actions of the individual elements of that sequence and the less predictable are future actions and necessities.

Corollary 2: The later in the development of any historical sequence, the more ordered the actions of the individual elements of that sequence and the more predictable are future actions and necessities.

Corollary 3: The actions of the individual elements of any historical sequence are generally postdictable but not specifically predictable, as regulated by corollaries 1 and 2.

Corollary 4: Change in historical sequences from chaotic to ordered is common, gradual, followed by relative stasis, and tends to occur at points where poorly established necessities give way to dominant ones so that a contingency will have little effect in altering the direction of the sequence.

Corollary 5: Change in historical sequences from ordered to chaotic is rare, sudden, followed by relative nonstasis, and tends to occur at points where previously well-established necessities have been challenged by others so that a contingency may push the sequence in one direction or the other.

At the beginning of any historical sequence, actions of the individual elements are chaotic, unpredictable, and have a powerful influence in the future development of that sequence. As the sequence gradually develops and the pathways slowly become more worn, out of chaos comes order. The individual elements sort themselves into their allotted positions, as dictated solely by what came before—the unique and characteristic sum and substance of history, driven forward on the entropic arrow of time by the interplay of contingency and necessity.

As a *model* of history (not a law), contingent-necessity provides a general cognitive framework for representing historical phenomena that does what the philosopher of science Rom Harré requires for models to be useful by distinguishing between the source and subject of a model: "The *subject* is, of course, whatever it is that the model represents. . . . The *source* is whatever it is the model is based upon."[13] When subject and source are the same, Harré calls these models homeomorphs; when different, paramorphs. A model airplane that represents both airplanes and a particular model of an airplane is a homeomorph model since source and subject are the same. Charles Darwin's model of the tree of life is a paramorph model since source and subject are different. The model of contingent-necessity is a paramorph model because, while as a general framework it may describe specific historical subjects, its source is not any one particular subject. Since the model attempts to explain a wide variety of historical examples, its source and subject are different.

As a *historical sequence*, the model of contingent-necessity is limited in its analysis to a specified range of chronological margins that are chosen by the historian. In other words, a historical sequence is what the historian says it is. The variables, however, are not arbitrarily chosen. The historian can present evidence for the significance of these precise starting and stopping points, which is what we already do. Once these chronological boundaries are established, then the model of contingent-necessity and its corollaries can be used as a heuristic framework for representing what happened between these limiting termini. *A historical sequence is a time frame determined by the focal point and the boundaries of the subject under investigation.*

Contingent-necessity and chaos are not dissimilar. At the 1986 Royal Society of London conference on chaos, Ian Stewart reports that scientists devised this seemingly paradoxical definition of chaos: "Stochastic behaviour occurring in a deterministic system." Stewart notes that since "stochastic behaviour is . . . lawless and irregular," and "deterministic behaviour is rule by exact and unbreakable law," then "chaos is 'lawless behaviour governed entirely by law,'" an obvious contradiction.[14] When chaos scientists use the word, however, they do not mean actual lawless and random behavior; they mean *apparent* chaos. The order is hidden from view by traditional methods of looking (e.g., linear mathematics). The apparently chaotic actions of the phenomena exhibit an interaction between the small, contingent events of a sequence and the large, necessitating laws of nature. When looked at in the light of contingent-necessity, it becomes possible to see that this is not a paradoxical usage and the following definition may be made: *Chaos is a conjuncture of events compelled to a certain course of action by constraining prior conditions.* Thus, "stochastic behaviour occurring in a deterministic system" is a similar description to "a conjuncture of events compelled to a certain course of action by constraining prior conditions." Chaos and contingent-necessity model phenomena in the same manner, as Ilya Prigogine notes when observing that in chaos the "mixture of necessity and chance constitutes the history of the system."[15] In like manner, necessity and contingency are the shaping forces for historical sequences. If this correspondence between chaos and contingent-necessity exists, then we can draw certain analogies between physical and biological chaotic phenomena and human historical ones.

According to Prigogine, all systems, including historical ones, contain subsystems that are "fluctuating." As long as the fluctuations remain modest and constant, relative stasis in the system is the norm. If the fluctuation becomes powerful enough that it upsets the preexisting organization and balance, a major change, or revolution, may occur, at which point the system may become chaotic. This point of sudden change is called the bifurcation point. Necessity takes a system down a certain path until it reaches a bifurcation point. At this time contingency plays an exaggerated role in nudging the system down a new path, which in time develops its own powerful necessities such that contingency is attenuated until the next bifurcation. It is the alloy of contingency and necessity that guides and controls the presence or absence of these bifurcations, as Prigogine notes in similar language: "The 'historical' path along which the system evolves as the control parameter grows is characterized by a succession of stable regions, where deterministic laws dominate, and of unstable ones, near the bifurcation points, where the system can 'choose' between or among more than one possible future."[16]

In the language of contingent-necessity, a bifurcation, or "trigger of change," is *any stimulus that causes a shift from the dominance of necessity and order to the dominance of contingency and chaos in a historical sequence.* Examples in history abound: inventions, discoveries, ideas, paradigm shifts, scientific revolutions, economic and political revolutions, war, famine and disease, invasions, immigrations and emigrations, population explosions, natural disasters, climate and the weather, and so on; all have the potential for triggering a sequence to change from order to chaos. A trigger of change, however, will not cause a shift at just any point in the sequence. Corollary 5 states that a trigger of change will be most effective when well-established necessities have been challenged by others so that a contingency may push the sequence in one direction or the other. This bifurcation point, or "trigger point," is *any point in a historical sequence where previously well-established necessities have been challenged by others so that a trigger of change (contingency) may push the sequence in one direction or the other.*

In like manner the sensitive dependence on initial conditions—the butterfly effect[17]—has direct application to corollaries 1 and 2 dealing with the point of time in the sequence—early or late—and the chaotic or ordered nature of that sequence. The butterfly effect, or the "trigger effect," is *the cascading consequences of a contingent trigger of change in a historical sequence.* The trigger effect is linked with corollary 1 where the earlier in the development of any historical sequence, the more chaotic the actions of the individual elements of that sequence and the less predictable are future actions and necessities, as well as with corollary 2, which reverses the influence of the trigger effect because the later in the development of any historical sequence, the more ordered the actions of the individual elements of that sequence, and the more predictable are future actions and necessities. Therefore the power of the trigger depends on *when* in the chronological sequence it enters. As stated in corollary 5, change tends to occur at points where previously well-established necessities have been challenged by others so that a contingency may push the sequence in one direction or the other. The flap of the butterfly's wings in Brazil may indeed set off a tornado in Texas, but only when the system has started anew or is precariously hanging in the balance. Once the storm is well under way, the flap of a billion butterfly wings would not alter the outcome for the tornado-leery Texans. The potency of the sequence grows over time. In human history the trigger effect is quickly erased once the patterns begin to settle in. An individual of great talent may have little effect in regions of stability, while another of modest competence might deflect the entire sequence in regions of instability. The "great men" of history found themselves at trigger points where well-established necessities had been challenged by others so that their contingent push jolted the sequence down a new path with the corresponding cascading consequences.

These corollaries do precisely what the philosopher George Reisch requires for laws to operate when they "divide the time over which its laws purportedly act into many small consecutive intervals or scenes. That is, covering-law explanations must be resolved into narrative temporal structures."[18] Historical sequences make up these consecutive (and contiguous) intervals over which the model of contingent-necessity operates. Corollaries 1 and 2 describe the chaotic or ordered nature of an interval depending on the temporal sequence within them; corollaries 4 and 5 describe when and why intervals shift from chaotic to ordered and vice versa. Whether these sequences are presented in the narrative or analytic form does not change the actions of the historical elements.

The QWERTY Principle of History

Regular users of typewriters and computers are locked by history into the standard QWERTY keyboard system (see figure 9.1), denoting the first six letters from the left on the top letter row. Our personal computer and typewriter keyboards still using the antiquated QWERTY system were designed for nineteenth-century typewriters whose key striking mechanisms were too slow for human finger speed. Even though more than 70 percent of English words can be produced with the letters DHIATEN-SOR, a quick glance at the keyboard will show that most of the letters are not in a strong striking position (home row struck by the strong first two fingers of each hand). Six of the ten letters are not on the home row (ITENOR are above and below) and one letter (A) is struck by the weak left little finger. All the vowels in QWERTY, in fact, are removed from the strongest striking positions, leaving only 32 percent of the typing on the home row. Only about one hundred words can be typed exclusively on the home row, while the weaker left hand is required to type over three thousand different words alone, not using the right hand at all. (It might also be noted that the word *typewriter* can be typed with letters all found on the top

Figure 9.1.   An early QWERTY keyboard

row. Apparently this was arranged so that typewriter salesmen could show off their new technology to prospective buyers with this clever trick.)[19]

This keyboard arrangement, in the early stages of its development, came about for numerous reasons and by a number of contingent causes; once set in motion (and given enough time), this conjuncture of events necessitated our inheritance of the system. With a check of the home row on the keyboard one can see the alphabetic sequence (minus the vowels) DFGHJKL. It would seem that the original key arrangement was just a straight alphabetical sequence, which makes sense in early experiments before testing was done to determine a faster alignment. But why remove the vowels? In Christopher Latham Sholes's original 1860s model, the QWERTY keyboard was designed to prevent key jamming because the paper was struck by the keys from underneath, and one could not see the page until it was nearly complete. Such an arrangement made it possible not only to type numerous unseen mistakes but to jam the keys all together and type a continuous line of one letter. The most used letters were removed from strong striking positions to slow the typist down. This problem was eventually remedied by using a front-facing roller with the paper scrolled around it so the typist could see each letter as it was struck. By then, however, QWERTY was so entrenched in the system (through manuals, teaching techniques, and other social necessities) that it became "locked in" or "path dependent." Once inevitable, the typewriter, along with the QWERTY keyboard, became entrenched in American business and culture.

In 1882, the Shorthand and Typewriter Institute in Cincinnati was founded by one Ms. Longley, who chose to adopt, among the many competing keyboard arrangements, the QWERTY system. As the school became well known her teaching methods became the industry standard, even adopted by Remington, which also began to set up typing schools using QWERTY. In 1888 a contest was arranged between Longley's method and that of her competitor Louis Taub, using a different typing technique on a non-QWERTY keyboard. Longley's star pupil, Frank McGurrin, apparently thrashed the competition by memorizing the entire QWERTY keyboard and typing by touch alone—an innovative technique for the time. The event generated much publicity and touch-typing became the method of choice for American typists. The Sholes-designed and Remington-manufactured typewriter with the QWERTY keyboard became necessary to the point that it would take a typing revolution to reshuffle the keyboard deck. Unless the major typewriter and computer companies, along with typing schools, teachers and publishers of typewriter manuals, and a majority of typists, all decided to change simultaneously, we are stuck with the QWERTY system indefinitely. An informal version of the model of contingent-necessity is what might be called the QWERTY principle: *historical events that come together in an unplanned way create inevitable historical outcomes.*

Beyond Chaos: Self-Organization, Antichaos, Simplexity, and Feedback

Sensitive dependency and bifurcations are only two aspects of chaos to be considered. "The straw that broke the camel's back" provides another metaphor of nonlinear dynamics. As the straw is piled on piece by piece, the camel's legs do not slowly bend lower until belly meets ground; rather, the camel maintains its straight-up stature until a critical breaking point is reached that causes the legs to suddenly buckle. Such is the nature of earthquakes, avalanches, economic depressions, ecological disasters, and quite possibly wars, revolutions, paradigm shifts, and other catastrophic historical events. Per Bak and Kan Chen describe this phenomenon as *self-organized criticality*, where "many composite systems naturally evolve to a critical state in which a minor event starts a chain reaction that can affect any number of elements in the system." They note that large systems such as geological plates, the stock market, and ecosystems "can break down not only under the force of a mighty blow but also at the drop of a pin." This one-pin-too-many, when dropped into a system in a critical or delicately balanced state, starts "a chain reaction that can lead to a catastrophe." Calling self-organized criticality a holistic theory, Bak and Chen claim: "Global features of the system cannot be understood by analyzing the parts separately. To our knowledge, self-organized criticality is the only model or mathematical description that has led to a holistic theory for dynamic systems."[20]

Their metaphorical model and experimental test of a catastrophe is a pile of sand in which single grains are dropped onto a flat, circular surface. The pile grows slowly and gradually into a gentle slope, and can be described by linear mathematics. But "now and then, when the slope becomes too steep somewhere on the pile, the grains slide down, causing a small avalanche."[21] This avalanche, or catastrophe, is triggered not by a major event or necessitating force, but by a minor contingency. (The avalanche, of course, could be triggered by a major event, but not necessarily.) The point at which the avalanche occurs Bak and Chen call chaotic. As the pile grows larger and larger, it "evolves on the border of chaos. This behavior, called weak chaos, is a result of self-organized criticality."[22] Because the actions and locations of the early falling grains affect the actions and locations of later falling grains, "the dynamics of a system are strongly influenced by past events."[23] In other words, history determines the system.

This action is what the model of contingent-necessity and its corollaries predict. The fall of the grains will be chaotic in the early stages and more ordered in the later, until reaching a critical point when a single grain can cause the whole system to collapse and start again. Contingencies (small grains of sand) construct necessities (large piles of sand), which grow to a point of criticality where one more contingent event (a single grain) can trigger a sudden and chaotic change. The model and its corollaries (particularly corollary 5), itself a holistic theory for dynamical systems, explain where and when certain grains make a significant difference while others do not.

Stuart Kauffman has examined the greatest organizing principle in all of biology—evolution—and argues that the large-scale origins and development of life may have come about naturally as a result of what he calls antichaos. Noting that chaos theory explains the randomizing force that causes systems to become disorderly, Kauffman reverses the principle to argue that "there is also a counterintuitive phenomenon that might be called antichaos: some very disordered systems spontaneously 'crystallize' into a high degree of order."[24] In the human genome, for example, out of the approximately 100,000 genes that code for the structure and function of a human body and brain, order naturally develops in the construction of approximately one hundred different cell types. Likewise for the evolution of life itself: "Organisms might have certain properties not because of selection but because of the self-organizing properties of those systems on which the selection works."[25] This emergence from chaos to order is described by corollaries 1 and 4 and is a normal function found in many physical, biological, and social phenomena.

What Kauffman calls antichaos, Cohen and Stewart call simplexity, or "the emergence of large-scale simplicities as a direct consequence of rules," or laws of nature.[26] These predictable laws interact with unpredictable contingencies to occasionally trigger the "collapse of chaos." After the collapse simple rules "emerge from underlying disorder and complexity," again, as described in corollaries 1 and 4. The antonym of Cohen and Stewart's simplexity is complicity, "the tendency of interacting systems to coevolve in a manner that changes both, leading to a growth of complexity from simple beginnings—complexity that is unpredictable in detail, but whose general course is comprehensible and foreseeable."[27] Complicity is a restatement of the model of contingent-necessity, especially corollary 3.

Calling their work the first postchaos, postcomplexity analysis, Cohen and Stewart argue that where chaos theory shows that simple causes can produce complex effects, and complexity theory reveals that complex causes can produce simple effects, they demonstrate that simplicity is generated from the interaction of chaos and complexity: "We argue that simplicities of form, function, or behavior emerge from complexities on lower levels because of the action of external constraints";[28] this is not unlike a conjuncture of events compelling a certain course of action by constraining prior conditions, or contingent-necessity.

Mass Hysterias as Chaotic Phenomena

To prove that history also exhibits properties of complexity, antichaos, self-organization, and feedback mechanisms, it will be demonstrated that certain historical phenomena repeat themselves, not in specifics but in universals. The witch crazes of the sixteenth and seventeenth centuries iterate many modern New Age social movements, specifically mass hysterias, moral panics, alien abduction claims, fear of Satanic cults, and the repressed memory movement. The parallels of these social movements to the witch crazes are too close to be accidental. Components of the witch craze movement are still alive in these modern descendants:

1. Women are usually the victims.
2. Sex or sexual abuse is often involved.
3. The mere accusation of potential perpetrators proves their guilt.
4. Denial of guilt is further proof of guilt.
5. Once victimization becomes well known in a community, others suddenly appear with similar claims.
6. The movement hits a critical peak of accusation where almost anyone can be a potential suspect and no one is above reproach.
7. The pendulum swings the other way as the innocent begin to fight back against their accusers and skeptics demonstrate the falsity of the claims.
8. Finally, the movement begins to fade, the public loses interest, and proponents are shifted to the margins of belief.

Why should there be such movements in the first place, and what drives these seemingly dissimilar systems in such a parallel manner? Stewart's self-organization and complexity explain how disordered systems become more and more complex until they reach a critical point upon which there is a dramatic change. Bak and Chen's self-organized criticality shows that at that point, when the system is in a critically balanced state, any stimulus may trigger a catastrophe. Not to be outdone in neologisms, John Casti calls this interactive process complexification, expressed by the interaction of two systems, such as an investor with the stock exchange, or with more general "feedback/feedforward loops" in the economy.[29] Feedback systems are those whose outputs are connected to their inputs in such a manner that there is constant change in response to both, much like microphone feedback in a PA system, or heart-rate monitoring in a biofeedback system. The rate of information exchange is the mechanism that fuels the feedback loop and drives growth to the point of criticality (e.g., stock prices go up and down in response to normal buying and selling; booms and busts are driven by a flurry of buying or selling when the system reaches criticality).

Social movements self-organize, grow, reach a peak, and then collapse, all described by corollaries 1–5. To be more descriptive, to the model of contingent-necessity I add here:

Corollary 6: Between origin and bifurcation, sequences self-organize through the interaction of contingencies and necessities in a feedback loop driven by the rate of information exchange.

In all of these social movements, from the witch crazes of centuries past to the New Age claims of today, one can find these nonlinear systems. The witch crazes are repeating themselves as these modern descendants because of the similarity of the components of the systems (the eight points above) and parallel social conditions in history, such as socio-economic stresses, cultural and political crises, religious and moral upheavals, the social control of one group of people by another in power, and a feeling of loss of personal control and responsibility such that an enemy is needed to blame. So it was for the witch crazes in England and New England (Salem), triggered by the social stresses of the sixteenth and seventeenth centuries. So it is for these New Age movements today, following sharply on the heels of the turbulent 1960s, the areligious and amoral 1980s, and the egalitarian and victimized 1990s.

These disordered social systems self-organize when a few claims are communicated to the society at large primarily through the oral tradition (in the sixteenth century) and the mass media (today). The feedback loop is then in place. A woman is accused of being a witch and is publicly burned. News of the event spreads by word-of-mouth. (A woman in therapy "remembers" being abused and accuses her father. Her story is published in a few journals and small newspapers.) Suddenly other women in the community are similarly accused and burned. Neighboring communities hear about it and begin their own purge. (Suddenly other women who read these stories of abuse enlist in therapy and recall their own repressed memories.) The system is now growing more complex. As the rate of information exchange increases, the positive feedback loop grows and the event snowballs into a full-blown social movement. Witches are held responsible for all of the community's woes and are accused, tried, and burned en masse. (Repressed memory is the subject of all the talk shows, magazine programs, tabloids, major newspapers, and self-help books. Women are accusing their alleged assailants en masse.) Figure 9.2 shows the growth and collapse of the witch craze sequence from 1560 to 1620. Figure 9.3 shows what drives the sequence from self-organization to criticality—the rate of information exchange in a feedback loop between suspected witches and their accusers (top) and between villages (bottom).[30] Figure 9.4 documents the number of accusations of sexual abuse against parents, officially registered by the False Memory Syndrome Foundation, from almost nothing in 1992 to almost twelve thousand cases in 1994.[31]

The movement grows in complexity to its level of criticality, whereupon it collapses. Hundreds of thousands of women (perhaps even a million) were killed and in some villages significant portions of the population were depleted until the psychosocioeconomic pressures were relieved and the witch craze collapsed. As for the repressed memory movement, the movement hit its peak in the mid-1990s and was largely dormant by the end of the century, triggered by a number of factors, including a successful settlement by Laura Pasley against her therapist, which has led to dozens of lawsuits filed against therapists; Gary Ramona's successful lawsuit and $500,000 award against his daughter's two therapists, opening the door to other third-party lawsuits; the plethora of "retractors"—in the past several years over three hundred women have retracted their original accusation; and perhaps most important in the reversal of the feedback loop, major media coverage now given to false memory syndrome rather than to a repressed memory syndrome.[32]

Alien abduction stories also nicely follow this sequence. Driven socially by cold war anxieties and the burgeoning science fiction industry, alien abduction claims took off in 1975 after millions watched NBC's portrayal of Betty Hill's abduction dreams as reality in *The UFO Incident*. The stereotypical alien with a bulbous bald head and large almond-shaped eyes, reported by so many abductees since 1975, was actually created by NBC artists. The rate of information exchange accelerated as more and more abductions were reported in the news. As there seemed to be corroboration on the appearance of the aliens, as well as the sexual content of the experiences (usually women being sexually molested in their beds by the aliens), the feedback loop was established. The sequence received a

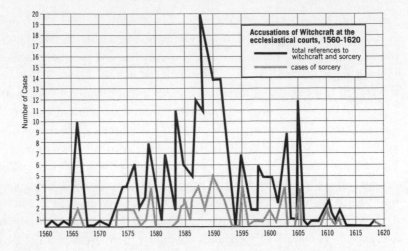

Figure 9.2.　The growth and collapse of the witch craze sequence from 1560 to 1620

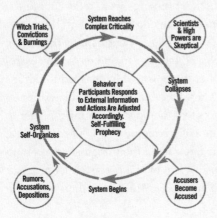

Figure 9.3 (*left*).　A witch craze feedback loop model. The driving force in the witch craze: the rate of information exchange between suspected witches and their accusers (top) and between villages (bottom).

Figure 9.4 (*below*).　A false memory sex abuse witch craze. The documented number of accusations of sexual abuse against parents, officially registered by the False Memory Syndrome Foundation, from almost nothing in 1992 to almost twelve thousand cases in 1994.

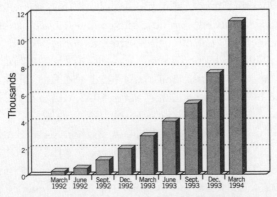

boost from academia when Harvard psychiatrist John Mack published his best-selling book *Abduction*, making the usual rounds of talk shows and, not surprisingly, causing an increase in abduction claims.[33]

Another classic example of mass hysteria is the strange case of the "phantom anesthetist" of Mattoon, Illinois, during the month of September 1944. On September 1, a woman claimed that a stranger entered her bedroom and anesthetized her legs with a spray gas. The *Mattoon Daily Journal-Gazette* ran the headline "Anesthetic Prowler on Loose." Soon cases were reported all over Mattoon, the state police were brought in, husbands stood guard with loaded guns, and dozens of reports of eyewitness sightings were made. After ten days, however, no one was caught, no chemical clues were discovered, the police spoke of wild "imaginations," and the newspapers began to report the story as a case of "mass hysteria." With this the movement reached criticality, the feedback loop reversed, and the last attack was reported on September 12. As described in corollary 6, the sequence had self-organized, reached complex criticality, switched from a positive to a negative feedback loop, and collapsed, all in the span of two weeks.[34]

The Pasteboard Masks of Laplace's Demon

In 1814 the French mathematician Pierre-Simon de Laplace published his *Essai Philosophique* in which he created the ultimate thought experiment: "Let us imagine an Intelligence who would know at a given instant of time all forces acting in nature and the position of all things of which the world consists; let us assume, further, that this Intelligence would be capable of subjecting all these data to mathematical analysis. Then it could derive a result that would embrace in one and the same formula the motion of the largest bodies in the universe and of the lightest atoms. Nothing would be uncertain for this Intelligence. The past and the future would be present to its eyes."[35]

Laplace's demon created a satanic scare that set the science of nonlinear dynamics back a century and a half. It was a chimera. The demon does not exist. But physicists chased the dream, and historians chased the physicists. Then the physicists created a new language, and historians learned to decipher the words. But the words bespoke an ancient language already known by the historians, and through a chaotic fog the physicists could hear Clio, the muse of history, ask Urania, the muse of astronomy: Is not the apparent chaos of history the fog through which we peer to make out a faint outline of meaning to chart our course from the past to the future? Is not the mystery of our past like the whiteness of Ahab's whale, possessing us to "chase him round Good Hope, and round the horn, and round the Norway Maelstrom, and round perdition's flames before I give him up"? Only in the pursuit of this hue "can we thus hope to light upon some chance clue to conduct us to the hidden cause we seek." How do we get to this hidden cause? Herman Melville seems to be telling us, through Captain Ahab's obsessive quest for "that accursed white whale that razeed me; made a poor pegging lubber of me for ever a day," that we must unmask the demon and penetrate the outer layers, no matter how obfuscating, with every tool available.

> All visible objects, man, are but as pasteboard masks. But in each event—in the living act, the undoubted deed—there, some unknown but still reasoning thing puts forth the mouldings of its features from behind the unreasoning mask. If man will strike, strike through the mask! How can the prisoner reach outside except by thrusting through the wall? To me, the white whale is that wall, shoved near to me. Sometimes I think there's naught beyond. But 'tis enough. He tasks me; he heaps me; I see in him outrageous strength, with an inscrutable malice sinewing it. That inscrutable thing is chiefly what I hate; and be the white whale agent, or be the white whale principal, I will wreak that hate upon him.[36]

Read history for whale and facts for masks.

The Tale of the Typewriter: A Postscript to Chapter 9

In "Exorcising LaPlace's Demon," I present what I call the QWERTY Principle: *Historical events that come together in an unplanned way create inevitable historical outcomes.* My principle is derived from a historical case study involving the development of the typewriter keyboard. The QWERTY keyboard that began life on Christopher Latham Sholes's 1868 typewriter, purchased and mass produced by Remington in 1873, and still employed today by nearly all computer users, was locked into use by historical momentum even though it is not necessarily the most efficient design. The Dvorak keyboard, patented by August Dvorak in 1936, for example, was claimed by Dvorak and others to be more efficient, and they advocated its use on typewriters that did not suffer from the key jamming problem that supposedly led to Sholes's keyboard (designed to slow typists down). It was not adopted, it is argued, because of historical "lock-in" and "path dependency": QWERTY was already so far down the path of historical market momentum that users would not make the switch.

I was inspired to derive my principle from an article by the late Harvard paleontologist Stephen Jay Gould, in an essay he wrote entitled "The Panda's Thumb of Technology." The panda's thumb was Gould's favorite example of the bottom-up tinkering of evolution, which uses whatever biological equipment is available for use, in opposition to the creationists' claim that nature shows top-down intelligent design. The panda does not have a thumb. The digit it uses to strip the leaves off plants is called its "thumb," but it is, in fact, an extended radial sesamoid (wrist) bone. It already has the traditional five digits, all locked into place through muscles, tendons, and nerves and designed for tearing and clawing from forward to back. Its extra digit is not an intelligently designed thumb; it is a jury-rigged wrist bone. The QWERTY keyboard, Gould argued, followed a similar principle of historical trajectory, its "design" a sign of its past development, not its current usage.

Since the time I conducted my research for this article, the claim that QWERTY serves as an example of suboptimal path dependency and historical lock-in has been challenged by several researchers. It is claimed that Sholes designed QWERTY to prevent jamming not by slowing down the typist but by separating keys whose typebar letters were close to each other underneath the typewriter carriage, which caused jamming (he moved the T and the H apart, for example). The QWERTY design was thus apparently a result of Sholes' study of letter pair frequencies, not an impediment to fast typists whose alacrity conflicted with the mechanical ability of typewriters of that time.

Further, it is claimed that numerous typing tests have shown no significant differences between QWERTY and the Dvorak keyboard, and, ironically, it appears that the original test in which the Dvorak keyboard beat QWERTY in a 1944 match-up was conducted by none other than August

Dvorak himself, making his data suspect because of a conflict of interest with his desire to market and sell his invention. Some authors have even argued that QWERTY's inefficiencies—unbalanced loads on left and right hands for the most commonly used keys, and an excess loading on the top row of commonly used keys—may be compensated by the forced hand alternation during typing. This may explain why there is little overall difference in efficiency, even though with the Dvorak keyboard about four hundred of the most common words can be typed from the home row, whereas with QWERTY that figure is only about one hundred.

In response to these critiques, author and investigative journalist Randy Cassingham has compiled the most comprehensive database on this subject, from which he disputes these claims and maintains the superiority of the Dvorak keyboard. See, for example, his book, *The Dvorak Keyboard* (available at http://www.dvorak-keyboard.com/, along with numerous articles and books on the subject), as well as his response to a widely distributed *Reason* magazine article at http://www.Dvorak-Keyboard. com/dvorak2.html. According to Cassingham:

> The Dvorak has the most-used consonants on the right side of the home row, and the vowels on the left side of the home row. Among other design features, it is set up to facilitate keying in a back-and-forth motion—(right hand, then left hand, then right, etc.). When the same hand has to be used for more than one letter in a row (e.g., the common t-h), it is designed not only to use different fingers when possible (to make keying quicker and easier), but also to progress from the outer fingers to the inner fingers ("inboard stroke flow")—it's easier to drum your fingers this way (try it on the table-top). The back-and-forth flow obviously makes typing quicker and easier: try typing the word "minimum" on the QWERTY keyboard, then look how you'd type it on Dvorak. The row above the top row contains the next-most-used letters and punctuation. Why? Because it is easier for your fingers to reach up on the keyboard than down. The least-used keys are on the bottom row. [See figure 9.5 below, from Cassingham's Web page.]
>
> If it's so good, why isn't everyone using it? Clearly, a good question. There are a lot of reasons for it, and they are too numerous and complicated to explain fully here (they are fully explained in the book *The Dvorak Keyboard*). Mainly, the QWERTY keyboard

Figure 9.5.   The Dvorak keyboard.

became entrenched in tradition. By the time Dr. Dvorak came up with his design, there were hundreds of thousands of typewriters, and it would have cost millions to convert them all over. The switch was just about to be made, but then World War II broke out, and the War Dept. ordered all typewriter keyboards be set to the most-common standard—QWERTY—and typewriter manufacturers retooled to produce small arms. By the end of the war, QWERTY was cast in concrete.

Regardless of the details and outcome of the QWERTY v. Dvorak debate, this does not effect the validity of the QWERTY Principle. Historical lock-in and path dependency are real phenomena. Note, for example, that we still use the "shift" key on our computer keyboards to produce a capital letter, even though there is no "carriage" inside the computer that needs shifting, and we still stroke the "return" key, even though there is no carriage to be returned to the start position. QWERTY got a historical head start over other keyboards, and it has proved good enough to maintain its market dominance. If Dvorak (or some future alternative) proves superior to QWERTY, it will have to be superior enough to overcome personal and cultural momentum. Technological systems, like biological ones, lock in their form and function according to a combination of efficacy and history. Optimal v. suboptimal is not the only deciding factor. The market triumph of VHS over superior Beta videotapes, and of cassette tapes over higher-quality eight-track tapes, for example, are additional examples of path dependency and lock-in, yet it has been argued that VHS tapes and audiocassette tapes allow for longer recording times, a

value customers may weigh in market purchase decisions. When efficiency and optimality are not the deciding factors, consumer habit and
market history may be.

History matters. The panda's thumb is good enough for stripping leaves
off of bamboo shoots (clumsy though it may be), and thus there was no
reason for natural selection to undertake a major overhaul of the panda's
paw. Although cultural selection weakens the path dependency hardwired
into biological systems driven by natural selection, historical momentum
is still a demonstrably powerful force in both evolution and history.

10

What If?

Contingencies and Counterfactuals:
What Might Have Been and What Had to Be

IN RAY BRADBURY'S 1952 novelette *A Sound of Thunder,* the story's hero, Mr. Eckels, arranges an unlikely itinerary through a most unusual travel company, whose advertising marquee reads:

Time Safari, Inc.
Safaris to Any Year in the Past
You Name the Animal
We Take You There
You Shoot It

Time Safari, Inc., has studied the past so carefully that the company knows precisely when and where a certain animal is going to die. With the omniscience of Laplace's demon (see chapter 9), Time Safari, Inc., takes its customers back to just moments before the animal is about to meet its natural demise, at which time the hunter can nab his game.

There is, however, one critical stipulation—you must stay on a carefully chosen path that prevents any alteration of the past, as the guide explains: "It floats six inches above the earth. Doesn't touch so much as one grass blade, flower, or tree. It's an anti-gravity metal. Its purpose is to keep you from touching this world of the past in any way. Stay on the path."

Don't go off it. I repeat. Don't go off. For any reason! If you fall off, there's a penalty. And don't shoot any animal we don't okay." The consequences of violating this rule are clear: "We don't want to change the future. We don't belong here in the past. A Time Machine is finicky business. Not knowing it, we might kill an important animal, a small bird, a roach, a flower even, thus destroying an important link in a growing species."

Every hunter's favorite target, of course, are the dinosaurs, and Eckels heads for the Cretaceous, steps out of the time machine, meanders down the path, and prepares to bag his preselected *Tyrannosaurus*. Startled by the size and ferocity of the monstrous creature, Eckels stumbles off the path and into the moss. Other hunters shoot and kill the T. rex just moments before a giant tree limb was about to crush it in the original time sequence. The hunters pile back into the time machine and return to the present, bemoaning the fact that they are probably going to be fined for Eckels's breach. "Who knows what he's done to time, to history," the guide groans.

When Eckels departed, the country was in the midst of a political election in which the moderate candidate—Keith—was victorious. Had the extremist candidate—Deutscher—won, he would have established an oppressive dictatorship. Exiting the machine Eckels and the others notice that things are not quite the same. The wall advertisement now reads:

> Tyme Sefari Inc.
> Sefaris Tu Any Yeer En The Past
> Yu Naim the Animall
> Wee Taekyuthair
> Yu Shoot Itt

In this alternate history, Deutscher won. Predating chaos theory by decades, Bradbury took the butterfly effect—the sensitive dependence on initial conditions—quite literally in his counterfactual denouement.

> Eckels felt himself fall into a chair. He fumbled crazily at the thick slime on his boots. He held up a clod of dirt, trembling "No, it can't be. Not a little thing like that. No!" Embedded in the mud, glistening green and gold and black, was a butterfly, very beautiful and very dead. "Not a little thing like that! Not a butterfly!" cried Eckels. It fell to the floor, an exquisite thing, a small thing that could upset

balances and knock down a line of small dominoes and then big dominoes and then gigantic dominoes, all down the years across time. Eckels' mind whirled. It couldn't change things. Killing one butterfly couldn't be that important! Could it?

Could it? Not likely, but if it did it would alter the historical sequence down a radically different path, tweaking the future a lot more than just alternative spellings and election outcomes. A more likely result would have been no English at all as a language, or no democratic elections, or even no *Homo sapiens*. More likely, however, large-scale necessities would have washed over and averaged out such relatively insignificant contingencies. The flap of a butterfly's wings, or that of ten billion butterfly wings, in most circumstances most of the time would have been overridden by a couple of thundering T. rexes.

History is a product of *contingencies* (what might have been) and *necessities* (what had to be), the effects of both dependent on the time and circumstances of the event in question and the particular historical path on which it falls. Because of this, historians should continue doing what they have always done—write narratives with all their unique details, contingent events, and necessitating social and historical forces, blended into a complex story with plot and characters.

But historians can and should (if we want history to be a science) formulate models, metaphors, and (some day) perhaps even mathematical equations to explain not only why things happened as they did, but why they seem to happen over and over in their own contingently unique but necessarily similar fashion. Can we learn something about the past, and even the present (for which history is written), from considering alternate historical time lines? Can we play "What if?" games of history to any benefit beyond cocktail party ruminations? We can. In fact, we already do.

Counterfactual "What If?" History

One mode of humor employs a ridiculous exaggeration of a mundane or prosaic idea. Of the many humorous skits in this genre written by the original *Saturday Night Live* crew from the 1970s, one of the funniest was to take the parlor game of "What if?" history to its reductio ad absurdum, such as "What if Napoleon had the atom bomb?" Well, what if he did? Presumably he would have dropped one on Blücher as he rallied his troops at Waterloo. We know this is ridiculous, of course (and thus the laugh is earned), but hyperbole does not equal superfluity (surely Blücher, too, would have the bomb, and an early form of mutual assured destruction would have prevented its employment by either side). In moderation we can play this game to great historical insight, and the process even has a technical name: counterfactual conditionals.

In logic, conditionals are statements in the form "if p then q," as in "if Blücher arrives then Napoleon loses," where q depends on p (and p is the antecedent since it comes before q). Counterfactual conditionals alter the factual nature of p, where p' is counter to the facts, thus altering its conditional element q into q'. Counterfactual conditionals are said to be modal in nature, that is, changing the antecedent changes the modality of the causal relationship between p and q from necessary (what had to be) to contingent (what might have been). Change p to p' and instead of q you may get q', as in "if Blücher does not arrive, Napoleon may win." In other words, in counterfactual conditionals the modal nature of the relationship between p and q changes from necessary to contingent.

Here we see why few students ever get past introductory logic. This philosophical mouthful becomes meaningful through real-world examples. In my first paper published on chaos theory I offered a counterfactual conditional whereby the South might have gained independence from the North during the Civil War at the Battle of Antietam.[1] General George B. McClellan caught a break when one of his soldiers stumbled into Robert E. Lee's battle plans in the infamous Order 191, wrapped in cigar paper and accidentally dropped in an open field. With Lee's plans in hand, the impossibly refractory and interminably sluggish McClellan was able to fight Lee to a draw, thwarting the latter's invasion plans. In the factual time line the conditional sequence is "if McClellan has Lee's plans (p) then the invasion is turned back (q)." From this, additional conditional series arise where "if q then r (the war continues), s (England does not recognize the South as a sovereign nation), t (the Northern blockade continues), u (the South's diminishing resources hinders their war effort), v (Lee is defeated at Gettysburg), w (Grant becomes the head Northern general), x (Sherman destroys everything in his path from Atlanta to Savannah), *y* (Lee surrenders to avoid the utter destruction of the South), and *z* (America remains a single nation)."

In my counterfactual conditional McClellan does not get Lee's plans, he is dealt a major defeat, and the invasion continues until the South earns the recognition from England as a sovereign nation, bringing the British navy to bear on the Northern blockade (in order to maintain open trading channels), thereby allowing the South to replenish her rapidly depleting resources and carry on the war until the North finally gives up. Here p' (McClellan does not get Lee's plans) leads to q' (Lee wins), with the modal cascading consequences of r' through z', and America ever after is divided into two nations, changing almost everything that has happened since. Whether I am right about this counterfactual is not relevant here (and in this "What if?" genre a sizable literature exists on the Civil War alone). More than just an example of the modal nature of counterfactuals, we see how they help us think about cause-and-effect relationships in historical sequences. We don't just want to know *how* the Civil War unfolded, we want to know *why*. Why questions are deeper than how questions, and require an appropriately deeper level of analysis.

Implicit in nearly all historical narratives are counterfactual "What if?" scenarios. One cannot help but ask, "What if McClellan had not received Lee's plans?" and therefore, "What if Lee had successfully invaded the North and obtained a peace treaty with the North?" (See figure 10.1.) The consequences are obvious and the counterfactually modal reasoning pervasive.

In another paper on chaos and history I demonstrated the highly contingent nature of the Holocaust in general and Auschwitz in particular, showing how this tragedy—far from inevitable (as "intentionalists" would have us believe with Hitler plotting the gassing of Jews as far back as the First World War)—need not have been (and thus my thesis supports the "functionalist" position that the Holocaust was a function of a number of contingent events).[2] In examining in detail the contingencies leading to Auschwitz,[3] I turned to Robert Jan Van Pelt's *Auschwitz*, a book filled with counterfactuals.[4] Through a chronology of blueprints and architectural designs of Auschwitz (what he calls "a site in search of a mission") Van Pelt shows that modern myths about the camp have erased the historical contingencies of its origin and development.

Figure 10.1. What if the South had won the Civil War? That counterfactual may hinge on a very specific question: What if George B. McClellan had not accidentally received Robert E. Lee's battle plans wrapped in cigar paper?

> Banished from the world of description, analysis, and conclusion, Auschwitz has become a myth in which the assumed universality of its impact obscures the contingencies of its beginning. The result is an account of "blissful clarity" in which there are no contradictions because *statements of fact are interpreted as explanations*; "things appear to mean something by themselves."[5]

In the emphasized clause Van Pelt is arguing counterfactually that treating historical facts as explanations of what was inevitable is a myth. He deconstructs this myth by unraveling the contingencies that constructed the necessity that became the Auschwitz we have come to know today. The problem is that we are trying to understand the early stages of Auschwitz by what now remains. The original intention of Auschwitz was quite different. "Auschwitz was not preordained to become the major site of the Holocaust. It acquired that role almost by accident, and even the fact that it became a site of mass murder at all was due more to the failure to achieve one goal than to the ambition to realize another."[6] Auschwitz became a killing machine because of a historical counterfactual—the failure to achieve its original goal. The focus on Auschwitz's final stage, in fact, has prevented us from explicating its contingent history, as well as how ordinary men became criminals.

> We think of it as a concentration camp enclosed on itself, separated from the rest of the world by "night" and "fog." This almost comfortable demonization relegates the camp and the events that transpired there to the realm of myth, distancing us from all too concrete historical reality, suppressing the local, regional, and national context of the greatest catastrophe Western civilization both permitted and endured, and obscuring the responsibility of the thousands of individuals who enacted this atrocity step by step. None of them was born to be a mass murderer, or an accomplice to mass murder. Each of them inched his way to iniquity.[7]

Yet all the while, says Van Pelt, "the extermination of the Jews was meant to be a transient phenomenon in the history of the camp." Plans were continued to convert the camp yet again after the war, but "that other future never materialized. Thus the name Auschwitz became synonymous with the Holocaust, and not with Himmler's model town."[8] (See figure 10.2.)

That "other future" is the counterfactual Auschwitz, the model (and modal) town that Auschwitz might have been were it not for the historical antecedents in our time line. It is almost impossible not to think counterfactually, as Van Pelt does in rewinding the time line back even further: "Founded by Germans in 1270, Auschwitz was lost to the Reich in 1457 and, as it went to Austria, almost returned in 1772. It would have had the history of any other border town, but for the late-nineteenth-century German obsession with what was called 'the German East.'"[9] *It would have . . . but for* is classic counterfactual reasoning, and when you start looking for such counterfactuals in works of history they are pervasive, as historian Johannes Bulhof noted: "Indeed, by far the most common use of counterfactuals and modal claims is about ordinary historical subjects and is found in ordinary historical texts, as opposed to the extraordinary and sometimes dumbfounding articles which are often a source of amusement rather than serious historical reflection."[10]

Figure 10.2. What if Auschwitz had not become a death camp? A paper trail of architectural plans documents the development of Auschwitz from a labor camp to house Russian prisoners of war for the purpose of building an ideal Nazi town into a death factory. That counterfactual may hinge on a very specific question: What if the Germans had successfully conquered the Soviet Union?

In the controversial book *Hitler's Willing Executioners*, for example, Daniel Goldhagen writes in the counterfactual mode in defense of his thesis that deep-seated German anti-Semitism created the Holocaust.

> While members of other national groups aided the Germans in their slaughter of Jews, the commission of the Holocaust was primarily a German undertaking. Non-Germans were not essential to the perpetration of the Holocaust, and they did not supply the drive and initiative that pushed it forward. To be sure, had the Germans not found European (especially eastern European) helpers, then the Holocaust would have unfolded somewhat differ-

ently, and the Germans would likely not have succeeded in killing as many Jews.[11]

Again, note the counterfactual conditional—"had the Germans not . . . then the Holocaust"—and the use of such modal thinking to infer causality: if non-Germans were not essential to the Holocaust, then Germans were. Even the oft-employed quip "No Hitler, No Holocaust" is a counterfactual conditional.

Counterfactuals and Science

Whether Goldhagen is right is not my primary concern here, although I think he is not because of another counterfactual condition: how to explain the exceptions—Germans who helped Jews, and non-Germans who aided in the Holocaust (especially French and Poles). This should be a quantifiable and testable hypothesis; unfortunately Goldhagen never attempts to do so. Like most historians, he resorts to amassing data in favor of his viewpoint without directly addressing the counterfactuals.

Thus, my prime concern is more in considering counterfactuals and contingencies as additional tools of a science of history that help us to tease apart causal vectors. As we have seen, historians already use them extensively, so what I am proposing is that we acknowledge their use and more consciously identify them in their causal (or countercausal) role.

Counterfactual thinking, in fact, is prevalent in science once you look for it. Scientists study systems of interacting elements—astronomers examine the movement of planets, stars, and galaxies; biologists record the complex web of an ecosystem; psychologists observe the interactions of people in a crowd. The component parts of these systems can be labeled, as in the conditional "if p then q." If a star shows a certain type of wobble then it may have a planetary body orbiting it. If a chemical is introduced into an ecosystem then certain organisms may disappear. If a crowd is of a certain size then an individual in the crowd will probably comply to mob psychology. These are all real examples of conditional statements for which counterfactuals help us test hypotheses. The astronomer suspects that stellar wobble means a planetary body is present because of the counterfactual observation that when such wobble is not present in other nearby objects, no other body is present. The biologist knows about the relationship between introduced chemicals and local extinction because of the counterfactual observation that when the chemical is removed the affected species return. The psychologist understands the correlation between group size and social compliance because of the counterfactual observation that when group sizes are smaller, less compliance occurs.

In some sciences, the counterfactual conditionals are directly testable in the form of experimental and control group comparisons, tested definitively through statistical tools. In other sciences the counterfactuals must be inferred, as in the search for extra-solar-system planets, where none have ever been directly observed. The use of inference in the physical and biological sciences is commonplace, so we should not disparage its application in the historical sciences, including human history. An astronomer or biologist setting up a conditional string of components labeled "if p then $q, r, s, t, u, v, w, x, y, z$" uses the counterfactual conditional "if p' then $q', r', s', t', u', v', w', x', y', z'$" no differently than does the historian. Newton's famous formula $F = MA$ (force equals mass times acceleration) is, in its essence, the conditional statement, "*If* a certain force is applied *then* an object will move as a function of its mass and acceleration," or, counterfactually, "An object's acceleration is dependent on its mass and the force applied to it."

Unfortunately history has not found its Newton and no such simple formulas exist for historians (or any of the social sciences, for that matter). Still, a good start may be found comparing conditionals and counterfactual conditionals in historical time lines. That is, we can conduct thought experiments in the "if p then q" mode, informed by real historical examples of similar events to see how they unfolded in the "if p' then q'" condition. This is what is known as the comparative method in historical science. What I am proposing is that when we compare, say, one culture or one time period to another to see how and why they differ, we are also rerunning the time line in a counterfactual experiment. We cannot literally rerun the time line, of course, but we can approximate it. To show how, I present two counterfactual "What ifs?" of history: What if *Homo sapiens* had gone extinct and Neanderthals survived? and What if there had been no agricultural revolution?

What If Neanderthals Won and We Lost?

I first started thinking about these questions when I wrote a review of Robert Wright's book *Nonzero: The Logic of Human Destiny*[12] for the *Los Angeles Times*.[13] It did not dawn on me at the time, but *Nonzero* is a study in counterfactual historical reasoning. Wright's thesis is that over billions of years of natural history, and over thousands of years of human history, there has been an increasing tendency toward the playing of nonzero games between organisms. This tendency has allowed more nonzero gamers to survive. (In zero-sum games the margin of victory for one is the margin of defeat for the other. If the Yankees beat the Mets 8–2, the Mets lose 2–8—where the margin of victory is +6 and the margin of defeat –6, summing to zero. In non-zero-sum games both players gain, as in an economic exchange where I win by purchasing your product and you win by receiving my money.)

Although competition between individuals and groups was common in both biological evolution and cultural history, Wright argues that symbiosis among organisms and cooperation among people have gradually displaced competition as the dominant form of interaction. Why? Because those who cooperated by playing nonzero games were more likely to survive and pass on their genes for cooperative behavior. And this process has been ongoing, "from the primordial soup to the World Wide Web," including and especially hominids. From the Paleolithic to today, human groups have evolved from bands of hundreds, to tribes of thousands, to chiefdoms of tens of thousands, to states of hundreds of thousands, to nations of millions (and one with a billion). This could not have happened through zero-sum exchanges alone. The hallmarks of humanity—language, tools, hunting, gathering, farming, writing, art, music, science, and technology—could not have come about through the actions of isolated zero-sum gamers.

Wright's counterfactual reasoning comes in at the book's leitmotif: non-zero-sumness has produced direction in evolution, and this directionality means that "the evolutionary process is subordinate to a larger purpose—a 'higher' purpose, you might even say."[14] That purpose is to be found, says Wright, in the fact that our existence was necessary and inevitable. Replay the time line of life over and over and "we" would appear again and again, "we" being an intelligent social species that carries its "social organization to planetary breadth." Therefore, Wright concludes: "Globalization, it seems to me, has been in the cards not just since the invention of the telegraph or the steamship, or even the written word or the wheel, but since the invention of life. All along, the relentless logic of non-zero-sumness has been pointing toward this age in which relations among nations are growing more non-zero-sum year by year."[15]

Is social globalization an inevitable necessity of the evolutionary process? If *Homo sapiens* had not filled this ineluctable position of global dominance, would one of the other hominids or great apes have done so? This is a counterfactual question: "If they [*Homo sapiens*] had died out, would they have been the last?"[16] No, Wright concludes. "If our own ancestors had died out around that time, it probably would have been at the hands of the Neanderthals, who could have then continued on their co-evolutionary ascent, unmolested by the likes of us." What if Neanderthals had also gone extinct? "I'd put my money on chimps. In fact, I suspect that they are already feeling some co-evolutionary push; if they're not quite on the escalator, they're in the vicinity." What if all the great apes had gone extinct? "Well, monkeys, though more distant human relatives than any apes, can be pretty impressive. Baboons are cleverly coalitional, and macaques are quite creative."[17] Wright continues in the "What if?" modal mode:

> What if somehow the entire primate branch had been nipped in the bud? Or suppose that the whole mammalian lineage had never truly flourished? For example, if the dinosaurs hadn't met their untimely death, mightn't all mammals still be rat-sized pests scurrying around underfoot? Actually, I doubt it, but as long as we're playing "What if," let's suppose the answer is yes. So what? Toward the end of the age of dinosaurs—just before they ran into their epoch-ending piece of bad luck—a number of advanced species had appeared, with brain-to-body ratios as high as those of some modern mammals. It now looks as if some of the smarter dinosaurs could stand up and use grasping forepaws. And some may have been warm-blooded and nurtured their young. Who knows? Give them another 100 million years and their offspring might be riding on jumbo jets.[18]

Maybe, but it's a stretch looking that far back. Let's examine just one relatively recent historical counterfactual: What if Neanderthals won and we lost? In our time line, Neanderthals went extinct between forty thousand and thirty thousand years ago. What if we ran an alternate time line where *Homo sapiens* went extinct and Neanderthals continued flourishing in Europe, Asia, and the Levant? Would some big brow-ridged, stooped-shouldered, hirsute hominid now be sitting here writing an essay about how his species was inevitable?

Consider the facts. Neanderthals split off from the common ancestor shared with us between 690,000 and 550,000 years ago, and they were in Europe at least 242,000 (and perhaps 300,000) years ago, giving them free reign there for a quarter of a million years. They had a cranial capacity just as large as ours (ranging from 1,245 to 1,740 cc, with an average of 1,520 cc compared to our average of 1,560 cc), were physically more robust than us with barrel chests and heavy muscles, and sported a reasonably complex kit of about sixty different tools.[19] On the surface, then, it certainly seems reasonable to argue that Neanderthals had a good shot at "becoming us."

But if we dig below the surface there is almost no evidence that Neanderthals would have ever "advanced" beyond where they were when they disappeared thirty thousand years ago. Even though theories in paleoanthropology change by the season as new evidence is uncovered and scientists scramble to revise or demolish old theories (or create new ones), there is near total agreement in the literature that Neanderthals were not on their way to becoming anything. They were perfectly well-adapted organisms for their environments.

This progressivist bias, in fact, is pervasive in nearly all evolutionary accounts and is directly challenged by counterfactual thinking. I once explained to my young daughter that polar bears are a good example of a transitional species between land and marine mammals, since they are well adapted for both land and marine environments. But this is not correct. Polar bears are not "becoming" marine mammals. They are not becoming anything. They are perfectly well adapted for doing just what they do. They may become marine mammals should, say, global warming melt the polar ice caps. Then again, they may just go extinct. In either case, there is no long-term drive for polar bears to progress to anything since evolution creates only immediate adaptations for local environments. The same applies to our hominid ancestors. Yet as I write this essay, The Learning Channel's otherwise superb series *Dawn of Man* continues the bias with every segment throughout the four hours, recounting the steps taken by hominids "to become us," "on the long road to us," "on its way to becoming human," "at the midpoint of human evolution," and so on.

Let's examine the evidence. Paleoanthropologist Richard Klein, in his comprehensive and authoritative work *The Human Career*, concludes that "the archaeological record shows that in virtually every detectable aspect—artifacts, site modification, ability to adapt to extreme environments, subsistence, and so forth—the Neanderthals were behaviorally inferior to their modern successors, and to judge from their distinctive morphology, this behavioral inferiority may have been rooted in their biological makeup."[20] Neanderthals had Europe to themselves for at least 200,000 years unrestrained by the presence of other hominids, yet their tools and culture are not only simpler than those of *Homo sapiens*, they show almost no sign of change at all, let alone progress toward social globalization. Paleoanthropologist Richard Leakey notes that Neanderthal tools "remained unchanged for more than 200,000 years—a technological stasis that seems to deny the workings of the fully human mind. Only when the Upper Paleolithic cultures burst onto the scene 35,000 years ago did innovation and arbitrary order become pervasive."[21]

Likewise, Neanderthal art objects are comparatively crude and there is much controversy over whether many of them were not the product of natural causes instead of artificial manipulation.[22] The most striking exception to this is the famous Neanderthal bone flute dated from between 40,000 to 80,000 years before present (BP), which some archaeologists speculate means that the maker was musical. Yet even Christopher Wills, who is a rare dissenting voice who rejects the inferiority of the Neanderthals, admits that it is entirely possible that the holes were naturally created by an animal gnawing on the bone, not by some Paleolithic instrument maker. And even though Wills argues that "recent important discoveries suggest that toward the end of their career, the Neanderthals might have progressed considerably in their technology," he has to confess that "it is not yet clear whether this happened because of contact with the Cro-Magnons and other more advanced peoples or whether they accomplished these advances without outside help."[23]

Probably the most dramatic claim for the Neanderthals' "humanity" is the burial of their dead that often included flowers strewn over carefully laid-out bodies in a fetal position. I even used this example in my book

How We Believe, on the origins of religion,[24] but new research is challenging this interpretation. Klein notes that graves "may have been dug simply to remove corpses from habitation areas" and that in sixteen of twenty of the best-documented burial sites "the bodies were tightly flexed (in near fetal position), which could imply a burial ritual or simply a desire to dig the smallest possible burial trench."[25] Paleoanthropologist Ian Tattersall agrees: "Even the occasional Neanderthal practice of burying the dead may have been simply a way of discouraging hyena incursions into their living spaces, or have a similar mundane explanation, for Neanderthal burials lack the 'grave goods' that would attest to ritual and belief in an afterlife."[26]

Much has been made about the possibility of Neanderthal language, that quintessential component of modern cognition. This is inferential science at best since soft brain tissue and vocal box structures do not fossilize. Inferences can be drawn from the hyoid bone that is part of the vocal box structure, as well as the shape of the basicranium, or the bottom of the skull. But the discovery of part of an apparent Neanderthal hyoid bone is inconclusive, says Tattersall: "However the hyoid argument works out, however, when you put the skull-base evidence together with what the archaeological record suggests about the capacities of the Neanderthals and their precursors, it's hard to avoid the conclusion that articulate language, as we recognize it today, is the sole province of fully modern humans."[27] As for the cranial structure, in mammals the bottom of the cranium is flat but in humans it is arched (related to how high up in the throat the larynx is located). In ancestral hominids the basicranium shows no arching in *australopithecines*, some in *Homo erectus*, and even more in archaic *Homo sapiens*. In Neanderthals, however, the arching largely disappears, evidence that does not bode well for theories about Neanderthal language, as Leakey concludes: "Judging by their basicrania, the Neanderthals had poorer verbal skills than other archaic sapiens that lived several hundred thousand years earlier. Basicranial flexion in Neanderthals was less advanced even than in *Homo erectus*."[28]

Leakey then speculates, counterfactually, what would have happened had our ancestors survived: "I conjecture that if, by some freak of nature, populations of *Homo habilis* and *Homo erectus* still existed, we would see in them gradations of referential language. The gap between us and the rest of nature would therefore be closed, by our own ancestors."[29] That "freak of nature" is the counterfactual contingency that in our time line allowed us to survive while no other hominids did, and thus Leakey concludes, *"Homo sapiens* did eventually evolve as a descendant of the first humans, but there was nothing inevitable about it."[30] Ian Tattersall also reasons in the counterfactual mode: "If you'd been around at any earlier stage of human evolution, with some knowledge of the past, you might have been able to predict with reasonable accuracy what might be coming up next. *Homo sapiens,* however, is emphatically not an organism that does what its predecessors did, only a little better; it's something very—and potentially very dangerously—different. Something extraordinary, if totally fortuitous, happened with the birth of our species."[31] (See figure 10.3.)

In fact, if we want to get recursive about this, the very process of thinking counterfactually—that is, the ability to ask "What if?" questions— spotlights the best candidate for the trigger of the Great Leap Forward in human evolution. As Tattersall suggests: "Language is not simply the medium by which we express our ideas and experiences to each other. Rather it is fundamental to the thought process itself. It involves categorizing and naming objects and sensations in the outer and inner worlds and making associations between resulting mental symbols. It is, in effect, impossible for us to conceive of thought in the absence of language, and it is the ability to form mental symbols that is the fount of our creativity, for only once we create such symbols can we recombine them and ask such questions as 'What if. . . . ?' "[32]

Had Neanderthals won and we lost, there is every reason to believe that they would still be living in a Stone Age culture of hunting, fishing, and gathering, roaming the hinterlands in small bands of a couple of dozen individuals, surviving in a world without towns and cities, without music and art, without science and technology . . . a world so different from our own that it is almost inconceivable. All because they never asked "What if?"

What If There Had Been No Agricultural Revolution?

Let's now jump back into our own time line where Neanderthals went extinct and we flourished. Between 35,000 and 13,000 years BP, tool kits became much more complex and varied, clothing covered near-naked bodies, sophisticated representational art adorned caves, bones and wood

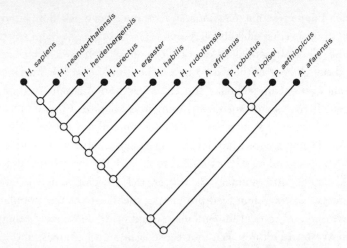

Figure 10.3. What if *Homo sapiens* had gone extinct and Neanderthals had survived? That counterfactual may hinge on a very specific question: What if *Homo sapiens* had never developed symbolic language? Without that there would have been no advance of complex tools, no art and music, no science and technology, no sophisticated culture, no towns and cities, no socially global dominant species.

formed the structure of living abodes, language produced symbolic communication, and anatomically modern humans began to wrap themselves in a blanket of crude but effective technology. They spread to nearly every region of the globe and all lived in a condition of hunting, fishing, and gathering (HFG). Some were nomadic, while others stayed in one place. Small bands grew into larger tribes, and with this shift possessions became valuable, rules of conduct grew more complex, and population numbers crept steadily upward. Then, at the end of the last Ice Age, roughly 13,000 years ago, population pressures in numerous places around the globe grew too intense for the HFG lifestyle to support. The result was the Neolithic Revolution, where the domestication of grains and large mammals produced the necessary calories to support the larger populations.[33]

But, counterfactually, what if those grains and mammals had not been available for exploitation? What if, by some quirk of biological evolution, of the tens of millions of species then living just fourteen mammals and a couple hundred plants had been erased from the biological record? What if we replayed the time line without those species critical for agriculture? What would our world look like today?

Such counterfactual questions, in fact, have been asked by Jared Diamond in his remarkable book *Guns, Germs, and Steel*, in which he explains the differential rates of development between civilizations around the globe over the past thirteen thousand years.[34] Why, Diamond asks, did Europeans colonize the Americas and Australia, rather than Native Americans and Australian Aborigines colonize Europe? Diamond rejects the theory that inherited differences in abilities between the races precluded some groups from developing as fast as others. Instead he proposes a biogeographical theory having to do with the availability of domesticated grains and animals to trigger the development of farming, metallurgy, writing, non-food-producing specialists, large populations, military and government bureaucracies, and other components that gave rise to Western cultures. Without these plants and animals, and a concatenation of other factors, none of these characteristics of our culture could exist. (See figure 10.4.)

How do we test this counterfactual argument? Through the comparative method. Compare, for example, Australia and Europe. Australian Aborigines could not strap a plow to or mount the back of a kangaroo, as Europeans did the ox and horse. Indigenous wild grains that could be domesticated were few in number and located only in certain regions of the globe—those regions that saw the rise of the first civilizations. The East-West–oriented axis of the Eurasian continent lent itself to diffusion of domesticated grains and animals as well as knowledge and ideas, so Europe was able to benefit much earlier from the domestication process.[35] By comparison, the North-South–oriented axis of the Americas, Africa, and the Asia-Malaysia-Australia corridor did not lend itself to such fluid transportation, and thus those areas already not well suited biogeographically for farming could not even benefit from diffusion. In addition, through constant interactions with domesticated animals and other peoples, Eurasians developed immunities to numerous diseases that, when brought by them in the form of germs to Australia and the Americas, along with their guns and steel, produced a genocide on a hitherto unseen scale.

An additional comparative test is seen in the fact that modern Australian Aborigines can learn, in less than a generation, to fly planes, operate computers, and do anything that any European inhabitant of Australia can do. Comparatively, when European farmers were transplanted to

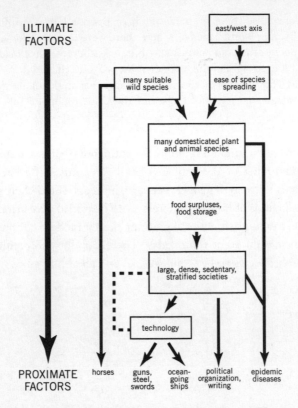

Figure 10.4. What if there had been no agricultural revolution? This counterfactual hinges on a specific question: What if the domesticable plants and animals around the world had not evolved? Paleolithic hunters and gatherers would not have been able to shift to farming, and thus they would not have been able to support large populations, without which there can be no division of labor, no cities, no developed science and technologies, and no civilization as we know it.

Greenland they went extinct when their environment changed, not because their genes prevented success.

In discussing my counterfactual examples with Diamond, he agreed with me about the nonprogressive fate of the Neanderthals, but demurred for agricultural contingency.

> Here I would argue for inevitability. The proof that it was inevitable was that it happened nine different times independently. If it hadn't happened in the Fertile Crescent 10,500 years ago, it would have

happened elsewhere, as, in fact, it did happen in China 9,500 years ago. Even in North America, where there were far fewer domesticable species, they did not just stay hunter-gatherers. They made the transition to agriculture and civilization, but they did so more slowly. They developed towns and villages, writing, metallurgy, and even mass production of tools, but they did so about 5,000 years later.[36]

Yet even agricultural inevitability is contingent upon the number, availability, and timing of domesticates. And, of course, if there were no domesticates thirteen thousand years ago, humans would likely still be living in small bands of hunter-gatherers, with beautiful cave art, interesting musical instruments, functional clothing, and a modest culture, but in a world radically different from today. There but for the contingency of domesticable species go I . . . and you . . . and all of humanity.

11

The New New Creationism
Intelligent Design Theory and Its Discontents

IN MARCH 2001 the Gallup News Service reported the results of their survey that found that 45 percent of Americans agree with the statement "God created human beings pretty much in their present form at one time within the last 10,000 years or so," while 37 percent preferred a blended belief that "human beings have developed over millions of years from less advanced forms of life, but God guided this process," and a paltry 12 percent accepted the standard scientific theory that "human beings have developed over millions of years from less advanced forms of life, but God had no part in this process."

In a forced choice between the "theory of creationism" and the "theory of evolution," 57 percent chose creationism against only 33 percent for evolution (10 percent said they were "unsure"). Only 33 percent of Americans think that the theory of evolution is "well supported by evidence," while slightly more (39 percent) believe that it is not well supported and that it is "just one of many theories." One reason for these disturbing results can be seen in the additional finding that only 34 percent of Americans consider themselves to be "very informed" about evolution. Clearly the 66 percent who do not consider themselves very informed about evolution have not withheld their judgment on the theory's veracity.

In any case, truth in science is not determined vox populi. It does not matter whether 99 percent or only 1 percent of the public believes a scientific theory—a scientific theory stands or falls on evidence, and there are few theories in science that are more robust than the theory of evolution. The preponderance of evidence from numerous converging lines of inquiry (geology, paleontology, zoology, botany, comparative anatomy, molecular biology, population genetics, biogeography, etc.) all independently converge to the same conclusion—evolution happened. The nineteenth-century philosopher of science William Whewell called this process a "consilience of inductions." I call it a "convergence of evidence." By whatever name, this is how historical events are proven.

The Evolution of Creationism

According to the first amendment of the United States Constitution, "Congress shall make no law respecting an establishment of religion, or prohibiting the free exercise thereof." How this applies to the creationism controversy over the past century evolved, as it were, through three stages, which together led to the birth of the new new creationism, Intelligent Design theory.

1. *The Banning of Evolution—That Old Time Religion.* In the 1920s, a perceived degeneration of the moral fiber of America was increasingly linked to Darwin's theory of evolution. In 1923, Oklahoma passed a bill offering free textbooks to public schools on the condition that evolution not be included. The same year Florida went even further by passing an antievolution law. In 1925, the Butler Act, making it "unlawful for any teacher in any of the Universities, Normals and all other public schools of the state . . . to teach any theory that denies the story of the Divine Creation of man as taught in the Bible, and to teach instead that man has descended from a lower order of animals," was passed by the Tennessee legislature. The bill was perceived to be in obvious violation of civil liberties and resulted in the famous Scopes trial. Despite a supposedly "moral" victory for Scopes, the controversy stirred by the trial made textbook publishers and state boards of education reluctant to deal with the theory of evolution in any manner, and the subject was simply dropped for decades until the *Sputnik* scare of 1957,

which rejuvenated science education. By 1961, the National Science Foundation, in conjunction with the Biological Science Curriculum Study, had outlined a basic program for teaching the theory of evolution and published a series of biology books whose common fiber was the theory.

2. *Equal Time for Genesis and Darwin—That Old Creationism.* The creationists responded with a new approach in which they demanded "equal time" for the Genesis story, along with the theory of evolution, and insisted that evolution was "only" a theory, not a fact, and should be designated as such. This strategy was challenged by scientists in many states, and was ultimately defeated in Arkansas. In 1965 Susan Epperson, a high-school biology teacher in Little Rock, filed suit against the state on the grounds that an antievolution bill passed in 1929 violated her rights to free speech. After her victory, the case was overturned by the Arkansas Supreme Court in 1967 and later appealed to the U.S. Supreme Court. In 1968 the Court found Epperson in the right and ruled that the law was "an attempt to blot out a particular theory because of its supposed conflict with the biblical account." On the basis of the Establishment Clause, the Arkansas law was interpreted as an attempt to establish a religious position in a public classroom and was therefore overturned. The Supreme Court ruled that all such antievolution laws were unconstitutional.

3. *Equal Time for Creation-Science and Evolution-Science—That New Creationism.* Since evolution could not be excluded from the classroom, and since the teaching of religious tenets was unconstitutional, the creationists invented "creation-science." Since academic honesty calls for a balanced treatment of competing ideas, they argued, creation-science should be taught side by side with evolution-science (note the clever parallel hyphenations). The creationists pressed state boards of education and textbook publishers to include the science of creation alongside the science of evolution. In 1981, Arkansas Act 590 was enacted into law by the governor, requiring "balanced treatment of creation-science and evolution-science in public schools; to protect academic freedom by providing student choice; to ensure freedom of religious exercise; to guarantee freedom of speech; . . . to bar discrimination on the basis of creationist or evolutionist belief." The constitutionality of Act 590 was challenged on May 27, 1981, with the filing of a suit by Reverend Bill McLean and others. The case was brought to trial in Little Rock on December 7, 1981, as *McLean v. Arkansas.* The contestants were, on one side, established

science, scholarly religion, and liberal teachers (backed by the ACLU) and, on the other, the Arkansas Board of Education and the creationists.

Federal judge William R. Overton of Arkansas ruled against the state on the following grounds: First, creation science conveys "an inescapable religiosity" and is therefore unconstitutional. "Every theologian who testified," Overton explained, "including defense witnesses, expressed the opinion that the statement referred to a supernatural creation which was performed by God." Second, the creationists employed a "two-model approach" in a "contrived dualism" that "assumes only two explanations for the origins of life and existence of man, plants and animals: It was either the work of a creator or it was not." In this either-or paradigm, the creationists claim that any evidence "which fails to support the theory of evolution is necessarily scientific evidence in support of creationism." But as Overton clarified in this summary, "Although the subject of origins of life is within the province of biology, the scientific community does not consider origins of life a part of evolutionary theory." Furthermore, "evolution does not presuppose the absence of a creator or God and the plain inference conveyed by Section 4 [of Act 590] is erroneous." Finally, Overton summarized the opinions of expert witnesses that creation science is not science, as the enterprise is usually defined: "A descriptive definition was said to be that science is what is 'accepted by the scientific community' and is 'what scientists do.'" Overton then listed the "essential characteristics" of science (as outlined by the expert witnesses, including evolutionary biologists Stephen Jay Gould and Francisco Ayala): "(1) It is guided by natural law; (2) It has to be explanatory by reference to natural law; (3) It is testable against the empirical world; (4) Its conclusions are tentative. . . ; and (5) It is falsifiable." Overton concluded: "Creation science as described in Section 4(a) fails to meet these essential characteristics."

The New New Creationism:
The Rise of Intelligent Design Theory

Out of the ashes of the Arkansas creation decision rose the phoenix of Intelligent Design (ID) theory. Realizing that even a hint of religiosity in their science would doom them to extinction, the new generation of creationists took the decision made by Overton seriously, and they began to focus solely on turning their religious beliefs into a genuine science—not just the transparent facade seen through by the courts but an actual scientific infrastructure that covertly supports an unspoken (and never to be spoken) religious faith.

Throughout the 1990s this new generation of creationists turned to "bottom up" strategies of hosting debates at colleges and universities, publishing books with mainstream academic and trade publishing houses, and enlisting the aid of academics like University of California, Berkeley, law professor Phillip Johnson and Lehigh University biochemist Michael Behe. In 1997, they even roped the conservative commentator William F. Buckley into hosting a PBS *Firing Line* debate, where it was resolved that "evolutionists should acknowledge creation." The debate was emblematic of the new creationism, employing new euphemisms such as "intelligent design theory," "abrupt appearance theory," and "initial complexity theory," where it is argued that the "irreducible complexity" of life proves it was created by an "intelligent designer." What does all this new language mean, and who are these guys anyway?

Intelligent Design Arguments and Rebuttals

The New New Creationists are nothing if not prolific. Their arguments can be found in a number of works published over the past decade, the most prominent and widely quoted of which include: William Dembski's *Intelligent Design* (InterVarsity Press), *No Free Lunch* (Rowman and Littlefield), and *The Design Inference* (Cambridge University Press); Phillip Johnson's *Darwin on Trial* (InterVarsity Press), *Reason in the Balance* (InterVarsity Press), and *The Wedge of Truth* (InterVarsity Press); *Darwin's Black Box* by Michael Behe (Simon and Schuster); *Darwinism, Design, and Public Education* edited by John Angus Campbell and Stephen C. Meyer; *The Creator and the Cosmos* and *The Fingerprint of God* (NavPress), both by Hugh Ross; *Of Pandas and People* by Dean Kenyon and William Davis (Haughton); *Evolution: A Theory in Crisis* by Michael Denton (Adler and Adler); *Icons of Evolution: Science or Myth?* by Jonathan Wells (Regnery).

A number of scientists began responding to the New New Creationism within a few years of the movement's rise to prominence. Kenneth Miller's

Finding Darwin's God (Perennial) and Robert Pennock's *Tower of Babel* (MIT Press) were the first two countershots that are indispensable in their analysis. Additional titles that should not be overlooked by those wishing a more in-depth analysis include: *Unintelligent Design* by Mark Perakh (Prometheus Books); *Creationism's Trojan Horse: The Wedge of Intelligent Design* by Barbara Forrest and Paul R. Gross (Oxford University Press); *God, the Devil, and Darwin: A Critique of Intelligent Design Theory* by Niall Shanks (Oxford University Press); *Darwin and Design: Does Evolution Have a Purpose?* by Michael Ruse (Harvard University Press); *A Devil's Chaplain* by Richard Dawkins (Houghton Mifflin); *Intelligent Design Creationism and Its Critics* edited by Robert Pennock (MIT Press); *Denying Evolution* by Massimo Pigliucci (Sinauer). Arthur Strahler's *Science and Earth History* (Prometheus) remains a classic, as do Richard Dawkins's *The Selfish Gene* and *The Blind Watchmaker*, and a number of Stephen Jay Gould's essay collections, such as *The Flamingo's Smile*. The two best resources on the Internet on the evolution/creation topic are the Talk Origins forum at www.talkorigins.org and Eugenie Scott's National Center for Science Education at http://www.natcenscied.org/.

Following the format of the "25 Creationists' Arguments and 25 Evolutionists' Answers" that I presented in my book *Why People Believe Weird Things*, we can review ID creationism in ten arguments and ten answers.

1. **The Nature of the Intelligent Designer.** *Many aspects of the universe and life indicate the fingerprint of intelligent design; thus an intelligent designer had a role in the creation of both the universe and of life. Since ID theory is a science, it cannot comment on the nature of this intelligent designer, let alone personalize it. The goal of ID theory is simply to establish the fact that the evidence is overwhelming that an intelligent designer was involved in the creation and evolution of the universe and life.*

The duplicity of the IDers is most apparent, and appalling, in their claim that they are only doing science and, therefore, they cannot comment on the nature of the intelligent designer. Why not? Are they not in the least bit curious as to who or what this ID is? If ID operates on the universe and our world, don't they want to know how ID works? They claim, for example, that certain biological and chemical systems are "irreducibly complex"—a number of different parts of a system could not possibly have come together by chance or through any other Darwinian or natural system or forces; therefore it must have happened through intelligent design. Granting, for the sake of argument, that they are right, if ID really did put together a number of biochemical components into a single cell in order to enable it to propel itself with a flagellum tail, or if ID did string together a number of molecules twisted into a double helix of DNA, don't ID theorists want to know *how* ID did it? Any scientist worth his or her sodium chloride would want to know. Did ID use known principles of chemical bonding and self-organization? If so, then ID appears indistinguishable from nature, and thus no supernatural explanation is called for; if not, then what forces did ID use?

In any case, is a set of natural laws and forces the sort of God whom IDers wish to worship? No. IDers want a *supernatural* God who uses unknown forces to create life. But what will IDers do when science discovers those natural forces, and the unknown becomes the known? If they join in the research on these mysteries then they will be doing science. If they continue to eschew all attempts to provide a naturalistic explanation for the phenomena under question, IDers will have abandoned science altogether. What a remarkably unscientific attitude. What an astounding lack of curiosity about the world. The British evolutionary biologist Richard Dawkins poignantly spelled this out in a clever fictional dialogue between two scientists. "Imagine a fictional conversation between two scientists working on a hard problem, say A. L. Hodgkin and A. F. Huxley who, in real life, won the Nobel Prize for their brilliant model of the nerve impulse," Dawkins begins.

"I say, Huxley, this is a terribly difficult problem. I can't see how the nerve impulse works, can you?"

"No, Hodgkin, I can't, and these differential equations are fiendishly hard to solve. Why don't we just give up and say that the nerve impulse propagates by Nervous Energy?"

"Excellent idea, Huxley, let's write the Letter to *Nature* now, it'll only take one line, then we can turn to something easier."

2. **Methodological Supernaturalism.** *Knowingly or unknowingly, scientists adhere to an underlying bias of methodological naturalism (sometimes called materialism or scientism), the belief that life is the result of a natural and purposeless process in a system of material causes and effects that does*

not allow, or need, the introduction of supernatural forces. University of California, Berkeley, law professor Phillip Johnson, a self-proclaimed "philosophical theist and a Christian" who believes in "a Creator who plays an active role in worldly affairs," claimed in his 1991 book Darwin on Trial *that scientists unfairly define God out of the picture by saying, essentially, "we are only going to examine natural causes and shall ignore any supernatural ones." This is limiting and restrictive. Theorists who postulate nonnatural or supernatural forces or interventions at work in the natural world are being pushed out of the scientific arena on the basis of nothing more than a fundamental rule of the game. Let's change the rules of the game to allow IDers to play.*

Okay, let's change the rules. Let's allow *methodological supernaturalism* into science. What would that look like? How would that work? What would we do with supernaturalism? According to ID theorists, they do not and will not comment on the nature of ID. They only wish to say, "ID did it." This reminds me of the Sidney Harris cartoon featuring the scientists at the chalkboard filled with equations, with an arrow pointing to a blank spot in the series denoting "Here a miracle happens." Although they eschew any such "god of the gaps"–style arguments, that is, in fact, precisely what they are doing. They have simply changed the name from GOD to ID.

For the sake of argument, however, let's assume that ID theorists have suddenly become curious about how ID operates. And let's say that we have determined that certain biological systems are irreducibly complex and intelligently designed. As ID scientists who are now given entrée into the scientific stadium with the new set of rules that allows supernaturalism, they call a time-out during the game to announce, "Here ID caused a miracle." What do we do now? Do we halt all future experiments? Do we continue our research and periodically say "Praise ID"? For the life of me I cannot imagine what we are supposed to do with *methodological supernaturalism* in the rules of the game of science.

There is, in fact, no such thing as the supernatural or the paranormal. There is only the natural, the normal, and mysteries we have yet to explain. It is also curious that ID miraculously intervenes just in the places where science has yet to offer a comprehensive explanation for a particular phenomenon. By a different name for a different time, ID (God) used to control the weather, but now that we have a science of meteorology ID has moved on to more obdurate problems, such as the origins of DNA or the evolution of cellular structures such as the flagellum. Once these problems are mastered then ID will presumably find even more intractable conundrums. Thus, IDers would have us teach students that when science cannot fully explain something we should look no further and declare that "ID did it." I fail to see how this is science. "ID did it" makes for a rather short lab lecture.

Finally, since ID creationists argue that what they are doing is no different from what the astronomers do who look for intelligent design in the background noise of the cosmos in their search for extraterrestrial intelligent radio signals (the SETI [Search for Extraterrestrial Intelligence] program, for example), then why not postulate that the design in irreducibly complex structures such as DNA is the result of an extraterrestrial experiment? Here is a viable hypothesis: ID = ET. Such theories have been proffered, in fact, by some daring astronomers and science fiction authors who speculated (wrongly it appears) that the earth was seeded with amino acids, protein chains, or microbes billions of years ago, possibly even by an extraterrestrial intelligence. Suffice it to say that no creationist worth his sacred salt is going to break bread or sip wine in the name of some experimental exobiologist from Vega. And that is the point. What we are really talking about here is not a scientific problem in the study of the origins of life; it is a religious problem in dealing with the findings of science.

3. **ID Intervention.** *According to the evidence, several billion years ago an Intelligent Designer created the first cell with the necessary genetic information to produce most of the irreducibly complex systems we see today. Then, the laws of nature and evolutionary change took over, and in some instances natural selection drove the system, except when totally new and more complex species needed creating. Then the Intelligent Designer stepped in again to intervene with a new design element.*

Just when and where ID intervened in the history of life is hotly disputed by ID theorists. Did ID trigger the big bang and laws of nature, then let the cosmos inflate and create its own subatomic and atomic particles? Or did ID do all of this, then let the stars create all of the other elements through natural processes? Did ID go so far as to generate all the stars and planets, along with the biochemical conditions necessary for life to arise, which then did on its own through natural forces? Or did ID take care of all of the physics and chemistry of life's creation and then let evolution take it from there? And as for the history of life itself, after (and however) it was created, did ID create each genus and then evolution create each species? Or did ID create each species and evolution create each sub-species? Most ID theorists accept natural selection as a viable explanation for microevolution—the beak of the finch, the neck of the giraffe, the varieties of subspecies found on earth. If ID created these species why not the subspecies? If natural selection can create subspecies, why not species? Or genus, for that matter? A species is defined as a group of actually or potentially interbreeding natural populations reproductively isolated from other such populations. We see evolution at work in nature today, isolating populations and creating new species, that is, new populations reproductively isolated from other such populations. If evolution can do this, why can't it also create higher-order categories of organisms? And if ID created the species, how did it do so? Did ID personally tinker with the DNA of every single organism in a population? Or did ID simply tweak the DNA of just one organism and then isolate that organism to start a new population? We are not told. Why? Because ID theorists have no idea and they know that if they want to be taken seriously as scientists they cannot just say, "ID did it."

An additional weakness in their argument can be seen in IDers' arrogant and indolent belief that if they cannot think of how nature could have created something through evolution, it must mean that scientists will not be able to do so either. This argument is not unlike those who, because they cannot think of how the ancient Egyptians built the pyramids, these structures must have been built by Atlantians or aliens. It is a remarkable confession of their own inabilities and lack of creativity. Who knows what breakthrough scientific discoveries await us next month or next year? The reason, in fact, that Michael Behe, author of *Darwin's Black Box*, has had to focus on the microscopic world's gaps is that the macroscopic gaps have mostly been filled. They are chasing science, not leading it. Also, sometimes we must simply live with uncertainties. A scientific theory need not account for every anomaly in order to be viable. This is called the residue problem—we will always have a "residue" of anomalies. It is certainly acceptable to challenge existing theories and call for an explanation of those anomalies. Indeed, this is routinely done in science. (The "gaps" that creationists focus on have all been identified by scientists first.) But it is not acceptable in science to offer as an alternative a nontestable, mystical, supernatural force to account for those anomalies.

Self-organization, emergence, and complexity theory form the basis of just one possible natural explanation for how the universe and life came to be the way they are. But even if this explanation turns out to be wanting, or flat-out wrong, what alternative do Intelligent Design theorists offer in its stead? If ID theory is really a science, as IDers claim it is, then the burden is on them to discover the mechanisms used by the Intelligent Designer. And if those mechanisms turn out to be natural forces, then no supernatural forces (ID) are necessary, and IDers can simply change their name to scientists.

4. **Irreducible Complexity.** *According to Lehigh University biochemist Michael Behe, in his book* Darwin's Black Box: *"By irreducibly complex I mean a single system composed of several well-matched, interacting parts that contribute to the basic function, wherein the removal of any one of the parts causes the system to effectively cease functioning." Consider the human eye, a very complex organ that is irreducibly complex—take out any one part and it will not work. How could natural selection have created the human eye when none of the individual parts themselves have any adaptive significance? Or consider the bacterial flagellum, the IDers' type specimen of irreducible complexity and intelligent design—it is not like a machine, it is a machine, and a complex one at that, without antecedents in nature from which it could have evolved in a gradual manner.*

There are a number of answers that refute this argument. Starting general, Michael Behe concludes his discussion of irreducible complexity by stating: "An irreducibly complex system cannot be produced directly (that is, by continuously improving the initial function, which continues to work by the same mechanism) by slight, successive modifications of a precursor system, because any precursor to an irreducibly complex system that is missing a part is by definition nonfunctional." Philosopher Robert Pennock, in his 1999 book *Tower of Babel*, noted that Behe here employs a classic fallacy of bait-and-switch logic—reasoning from something that is true "by definition" to something that is proved through empirical evidence. This is not allowed in the rules of reasoning.

Evolutionary biologist Jerry Coyne, in a review of Behe's book in *Nature* (September 1996), explained that biochemical pathways such as those claimed by Behe to be impossible to explain without an intelligent designer, "did not evolve by the sequential addition of steps to pathways that became functional only at the end," as Behe argues. "Instead, they have been rigged up with pieces co-opted from other pathways, duplicated genes, and early multi-functional enzymes." Behe, for example, claims that the blood-clotting process could not have come about through gradual evolution. Coyne shows that, in fact, thrombin "is one of the key proteins in blood clotting, but also acts in cell division, and is related to the digestive enzyme trypsin."

This is the same answer given to the nineteenth-century antievolution argument that wings could not have evolved gradually—of what use is half a wing? The answer is that the incipient stages in wing development had uses other than for aerodynamic flight; in other words, half wings were not poorly developed wings, they were well-developed something elses. Likewise with the incipient stages in the evolution of blood clotting, the flagellum motor, and the other structures claimed by IDers to be inexplicable through evolutionary theory. The principle can be illustrated simply in figure 11.1.

As for the human eye, it is not true that it is irreducibly complex, where the removal of any part results in blindness. Any form of light detection is better than none—lots of people are visually impaired with any number of different diseases and injuries to the eyes, yet they are able to utilize their restricted visual capacity to some degree and would certainly prefer this to blindness. No one asks for partial vision, but if that is what you get, then like all life-forms throughout natural history you learn to cope in order to survive.

There is a deeper answer to the example of the evolution of the eye, and that is that natural selection did not create the human eye out of a warehouse of used parts lying around with nothing to do, any more than Boeing created the 747 without the ten million halting jerks and starts beginning with the Wright Brothers. Natural selection simply does not work that way. The human eye is the result of a long and complex pathway that goes back hundreds of millions of years: a simple eyespot where a handful of light-sensitive cells provides information to the organism about an important source of the light—the sun; a recessed eyespot where a small surface indentation filled with light-sensitive cells provides addi-

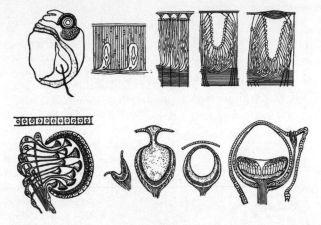

Figure 11.1.   Co-opting nature. At each stage of an evolutionary sequence a partic-ular structure, or series of structures, such as the eye, may serve one function, only to be co-opted later for some other use. The end product may appear to be designed for that final function, but it was not because evolution does not look ahead to the future.

tional data in the form of direction; a deep-recession eyespot where addi-tional cells at greater depth provide more accurate information about the environment; a pinhole camera eye that is actually able to focus an image on the back of a deeply recessed layer of light-sensitive cells; a pinhole lens eye that is actually able to focus the image; a complex eye found in such modern mammals as humans.

We should also note that the world is not always so intelligently designed, and the human eye is a prime example. The configuration of the retina is in three layers, with the light-sensitive rods and cones at the bot-tom, facing away from the light, and underneath a layer of bipolar, hori-zontal, and amacrine cells, themselves underneath a layer of ganglion cells that help carry the transduced light signal from the eye to the brain in the form of neural impulses. And this entire structure sits beneath a layer of blood vessels. (See figure 11.2.) For optimal vision, why would an intelli-gent designer have built an eye backward and upside down? This "design" makes sense only if natural selection built eyes from whatever materials were available, and in the particular configuration of the ancestral organ-ism's preexisting organic structures. The eye shows the pathways of evolu-tionary history, not intelligent design creation.

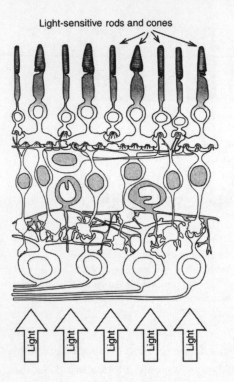

Light-sensitive rods and cones

Light Light Light Light Light

Figure 11.2. The poorly designed human eye. The anatomy of the human eye shows that it is anything but "intelligently designed." It is built upside down and backward, with photons of light having to travel through the cornea, lens, aqueous fluid, blood vessels, ganglion cells, amacrine cells, horizontal cells, and bipolar cells, before reaching the light-sensitive rods and cones that will transduce the light signal into neural impulses, where they are then sent to the visual cortex at the back of the brain for processing into meaningful patterns.

Finally, the bacterial flagellum, although a remarkable structure (figure 11.3), comes in many varieties of complexity and functions. In fact, bacteria in general may be subdivided into eubacteria and archaebacteria; the former are more complex and have more complicated flagella, while the latter are simpler and have correspondingly simpler flagella. Eubacterial flagellum, consisting of a three-part motor, shaft, and propeller system, is actually a more complicated version of the archaebacterial flagellum, which has a motor and a combined shaft-propeller system. So, when IDers describe the three-part flagellum as being irreducibly complex, they are wrong. It can be reduced to two parts. Additionally, the eubacterial flagellum turns out to be one of a variety of ways that bacteria move about their environment. Finally, the flagellum has functions other than just for propulsion. For example, for many types of bacteria the primary function of the flagellum is secretion, not propulsion. For others, the flagellum is used for attaching to surfaces and other cells. As for the evolution of the flagellum, we know that between eighteen and twenty genes are involved

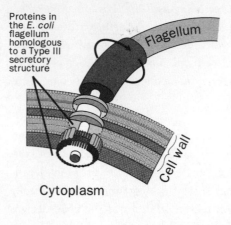

Proteins in the *E. coli* flagellum homologous to a Type III secretory structure

Flagellum

Cell wall

Cytoplasm

Figure 11.3. The bacterial flagellum: evolution at work. Long a favorite among IDers as an example of irreducible complexity and intelligent design, it turns out that there are a number of different types of bacterial flagella, ranging in complexity as well as serving a number of different functions, not just propulsion. Evolution well explains how, for example, the flagellum may have originally evolved as a mechanism for secretion, then later co-opted for propulsion.

in the development of the simpler two-part flagellum, twenty-seven genes make up the slightly more complex *Campylobacter jejuni* flagellum, and forty-four genes make up the still more complicated *E-coli* flagellum, a smooth genetic rise in complexity corresponding to the end product. And phylogenetic studies on flagella indicate that the more modern and complex systems share common ancestors with the simpler forms. So here an evolutionary scenario presents itself—archaebacteria flagella were primarily used for secretion, although some forms were co-opted for adhesion or propulsion. With the evolution of more complicated eubacteria, flagellum grew more complex, refining, for example, the two-part motor and shaft-propeller system into a three-part motor, shaft, and propeller system, which was then co-opted for more efficient propulsion.

5. **Inference to Design.** *In a special issue of the Christian magazine* Touchstone, *dedicated to Intelligent Design, Whitworth College philosopher Stephen Meyer argues that ID is not simply a "God of the gaps" argument to fill in where science has yet to give us a satisfactory answer—it is not just a matter of "we don't understand this so God must have done it." ID theorists like Meyer and Phillip Johnson, William Dembski, Michael Behe, and Paul Nelson (all leading IDers and contributors to this issue) say they believe in ID because the universe really does appear to be designed. "Design theorists infer a prior intelligent cause based upon present knowledge of cause-and-effect relationships," Meyer writes. "Inferences to design thus employ the*

standard uniformitarian method of reasoning used in all historical sciences,
many of which routinely detect intelligent causes. Intelligent agents have
unique causal powers that nature does not. When we observe effects that
we know only agents can produce, we rightly infer the presence of a prior
intelligence even if we did not observe the action of the particular agent
responsible."

Indeed, Psalm 19:1 declares: "The heavens declare the glory of God; and
the firmament showeth his handiwork." The design inference is not confined to the ancient Hebrews. In fact, in 1999 social scientist Frank J. Sulloway and I conducted a national survey, asking Americans why they
believe in God. The most common reason offered was the good design,
natural beauty, and complexity of the world. One subject wrote: "To say
that the universe was created by the Big Bang theory is to say that you can
create *Webster's Dictionary* by throwing a bomb in a printing shop and the
resulting explosion results in the dictionary."

The reason people think that a designer created the world is because,
well, it looks designed, and some evolutionary theorists, such as the
philosopher of science Michael Ruse, think that it's high time we quit tip-
toeing around this inference. In his book *Darwin and Design*, Ruse says we
should admit from the start that life looks designed because it was . . . from
the bottom up by evolution. Purpose follows functional adaptation: "At the
heart of modern evolutionary biology is the metaphor of design, and for this
reason function-talk is appropriate. Organisms give the appearance of
being designed, and thanks to Charles Darwin's discovery of natural selection we know why this is true. Natural selection produces artifact-like features, not by chance but because if they were not artifact-like they would
not work and serve their possessors' needs." More cautious evolutionary
theorists such as Ernst Mayr worry that "the use of terms like *purposive* or
goal-directed seemed to imply the transfer of human qualities, such as
intent, purpose, planning, deliberation, or consciousness, to organic structures and to subhuman forms of life." To which Ruse replies: "Well, yes it
does!" So what? At the heart of science is metaphor—Ruse notes that
physicists talk of force, pressure, attraction, repulsion, work, charm, and
resistance, all quite useful metaphors—and the metaphors of design and
purpose work well as long as we stick to natural explanations for nature

and understand that natural selection (another metaphor) is the primary mechanism for generating design and purpose, from the bottom up.

What role, then, is there for a top-down designer? If you are one of those 37 percent in the 2001 Gallup poll who believe that God guided the process of evolution then, on one level, you are in good company. In his 1996 encyclical *Truth Cannot Contradict Truth*, Pope John Paul II told a billion Catholics that, in essence, evolution happened—deal with it: "It is indeed remarkable that this theory has been progressively accepted by researchers, following a series of discoveries in various fields of knowledge. The convergence, neither sought nor fabricated, of the results of work that was conducted independently is in itself a significant argument in favor of the theory." Since both the Bible and the theory of evolution are true (and "truth cannot contradict truth"), John Paul II reconciled theological dualism with scientific monism by arguing that evolution produced our bodies while God granted us our souls.

This conciliatory position is fine as far as it goes, but many thinkers are not content to keep the magisteria of science and religion (per Gould) separate. They want empirical data to prove faith tenets, and it is here where the New New Creationism becomes William Paley redux. Paley was the eighteenth-century natural theologian whose "watchmaker" argument became the foundation of all modern design arguments. IDers recast Paley in modern jargon with new and more sophisticated biological examples (such as bacterial flagellum and blood-clotting agents). But as Darwin showed—and a century and a half of research has proven—the designer is a blind watchmaker (in Richard Dawkins's apposite phrase). Complex structures can and do arise out of simple systems through blind variation, selection, and adaptation. This is an inevitable outcome of Darwinism, says Ruse: "Whether we like it or not, we are stuck with it. The Darwinian revolution is over, and Darwin won." As pattern-seeking, storytelling primates who need origin myths, the theory of evolution now fulfills that need for us and has the added advantage that, unlike most origin myths, it is very probably true.

6. **Fine-Tuned Universe and Life.** *Physicist Freeman Dyson won the Templeton Prize valued at $948,000 for such works as* Disturbing the Universe, *one passage of which is very revealing: "As we look out into the*

universe and identify the many accidents of physics and astronomy that have
worked to our benefit, it almost seems as if the universe must in some sense
have known that we were coming." Mathematical physicist Paul Davies also
won the Templeton Prize, and we can understand why in such passages as
this from his 1999 book The Fifth Miracle:

> In claiming that water means life, NASA scientists are . . . mak-
> ing—tacitly—a huge and profound assumption about the nature of
> nature. They are saying, in effect, that the laws of the universe are
> cunningly contrived to coax life into being against the raw odds;
> that the mathematical principles of physics, in their elegant sim-
> plicity, somehow know in advance about life and its vast complexity.
> If life follows from [primordial] soup with causal dependability, the
> laws of nature encode a hidden subtext, a cosmic imperative, which
> tells them: "Make life!" And, through life, its by-products: mind,
> knowledge, understanding. It means that the laws of the universe
> have engineered their own comprehension. This is a breathtaking
> vision of nature, magnificent and uplifting in its majestic sweep. I
> hope it is correct. It would be wonderful if it were correct.

Such statements are powerful indeed, especially when uttered by prominent scientists not affiliated with the ID movement in any way. Even an
atheist like Stephen Hawking occasionally makes statements seemingly
supportive of scientistic arguments for God's existence:

> Why is the universe so close to the dividing line between collapsing
> again and expanding indefinitely? In order to be as close as we are
> now, the rate of expansion early on had to be chosen fantastically
> accurately. If the rate of expansion one second after the big bang
> had been less by one part in 10^{10}, the universe would have col-
> lapsed after a few million years. If it had been greater by one part in
> 10^{10}, the universe would have been essentially empty after a few
> million years. In neither case would it have lasted long enough for
> life to develop. Thus one either has to appeal to the anthropic prin-
> ciple or find some physical explanation of why the universe is the
> way it is.

That explanation, at the moment, is a combination of a number of different concepts revolutionizing our understanding of evolution, life, and
cosmos, including the possibility that our universe is not the only one. We

may live in a multiverse in which our universe is just one of many bubble universes, all with different laws of nature. Those with physical parameters like ours are more likely to generate life than others. But why should any universe generate life at all, and how could any universe do so without an intelligent designer?

The answer can be found in the properties of *self-organization* and *emergence* that arise out of what are known as complex adaptive systems, or complex systems that grow and learn as they change. Water is an emergent property of a particular arrangement of hydrogen and oxygen molecules, just as consciousness is a self-organized emergent property of billions of neurons. The entire evolution of life can be explained through these principles. Complex life, for example, is an emergent property of simple life: simple prokaryote cells self-organized to become more complex units called eukaryote cells (those little organelles inside cells you had to memorize in beginning biology were once self-contained independent cells); some of these eukaryote cells self-organized into multicellular organisms; some of these multicellular organisms self-organized into such cooperative ventures as colonies and social units. And so forth. We can even think of self-organization as an emergent property, and emergence as a form of self-organization. How recursive. No Intelligent Designer made these things happen. They happened on their own.

As a complex adaptive system the cosmos intelligently designs itself. It is one giant autocatalytic (self-driving) feedback loop that generates emergent properties, one of which is life. There may even be a type of natural selection at work among the many bubble universes, with those whose parameters are like ours most likely to survive. Those bubble universes whose parameters are most likely to give rise to life occasionally generate complex life with brains big enough to achieve consciousness and to conceive of such concepts as God and cosmology and to ask such questions as Why?

7. **Explanatory Gaps.** *ID theory fills in an explanatory gap that science cannot or has not filled. It is legitimate to identify the shortcomings of evolutionary theory, and show how scientists have not, and perhaps cannot, provide examples of evolution at work. It is one thing to infer in the fossil record a speciation event or the creation of a new structure; it is quite another to witness it in the laboratory. It is fair and reasonable to argue that students should be made aware of these explanatory shortcomings on the part of science.*

Not only does science have the incredibly rich fossil record, the process of evolution can be seen at work at a number of different levels. We know from genetics that every dog on the planet descended from a single population of wolves in China about 15,000 years ago. Granted, this was a combination of natural selection and artificial selection (breeding), but it also now appears from both genetics and paleontology that every human on the planet descended from a single population of *Homo sapiens* in Africa about 150,000 years ago. That's a lot of evolution in a relatively short period of time. And, of course, diseases are prime examples of natural selection and evolution at work, and on timescales we can witness, all too painfully. The AIDS virus, for example, continues to evolve in response to the drugs used to combat it—the few surviving strains of the virus continue to multiply, passing on their drug-resistant genes. This is evolution in action, which was even caught in a laboratory experiment published in the February 20, 2004, edition of *Science*, in which *E. coli* bacteria that were forced to adapt or perish improvised a novel molecular tool. According to the experimenter, University of Michigan biologist James Bardwell, "The bacteria reached for a tool that they had, and made it do something it doesn't normally do. We caught evolution in the act of making a big step." The big step was a new way of making molecular bolts called disulfide bonds, which are stiffening struts in proteins that also help the proteins fold into their proper, functional, three-dimensional shapes. This new method restarted the bacteria's motor and enabled it to move toward food before it starved to death.

This is an important experiment because Bardwell had developed a strain of mutant bacteria unable to make disulfide bonds, which are critical for the ability of the bacteria's propeller-like swimming motor, the flagella, to work. This is the same flagella that creationists are so fond of displaying as examples of irreducible complexity. The researchers put these nonswimming bacteria to the test by placing them on a dish of food where, once they had exhausted the food they could reach, they had to either repair the broken motor or starve to death. The bacteria used in the experiment were forced to use a protein called thioredoxin, which normally destroys disulfide bonds, to make the bonds instead. In a process similar to natural selection, one researcher made random alterations in the DNA-encoding thioredoxin and then subjected thousands of bacteria to the swim-or-starve test. He wanted to see if an altered version of thioredoxin could be coerced to make disulfides for other proteins in the bacteria. Remarkably, a mutant carrying only two amino acid changes, amounting to less than 2 percent of the total number of amino acids in thioredoxin, restored the ability of the bacteria to move. The altered thioredoxin was able to carry out disulfide bond formation in numerous other bacterial proteins all by itself, without relying on any of the components of the natural disulfide bond pathway. The mutant bacteria managed to solve the problem in time and swim away from starvation and multiply.

Bardwell concluded: "The naturally occurring enzymes involved in disulfide bond formation are a biological pathway whose main features are the same from bacteria to man. People often speak of Computer Assisted Design (CAD), where you try things out on a computer screen before you manufacture them. We put the bacteria we were working on under a strong genetic selection, like what can happen in evolution, and the bacteria came up with a completely new answer to the problem of how to form disulfide bonds. I think we can now talk about Genetic Assisted Design (GAD)."

Perhaps we should now talk about GAD instead of GOD.

8. The Conservation of Information and the Explanatory Filter.
According to William Dembski, information cannot be created by either natural processes or chance, so there is a law of conservation of information, which indicates design. Furthermore, design can be inferred through what Dembski calls an explanatory filter in the following way:

1. *If an event E has high probability, accept Regularity as an explanation; otherwise move to the next step.*
2. *If the Chance hypothesis assigns E a high probability or E is not specified, then accept Chance; otherwise move down the list.*
3. *Having eliminated Regularity and Chance, accept Design.*

First, Dembski's "Law of the Conservation of Information" is purposefully constructed to resemble such physical laws as the conservation of momentum or the laws of thermodynamics. But these laws were based on copious empirical data and experimental results, not inferred from logical argument as Dembski's law is. Second, no other recognized theory of information—such as those by Shannon, Kolmogorov, or Chaitin—includes a law or principle of conservation, and no one working in the information sciences uses or recognizes Dembski's law as legitimate, regardless of its design inference. Third, even if the law of the conservation of information were validated, it is irrelevant to the theory of evolution, because it is abundantly clear that information in the natural world—through DNA, for example—is transferred by natural processes. Fourth, the most common form of biological information transfer, DNA, in fact, has all the elements of historical contingency and evolutionary history, not design, as pointed out by evolutionary biologist Kenneth Miller.

> In fact, the human genome is littered with pseudogenes, gene fragments, "orphaned" genes, "junk" DNA, and so many repeated copies of pointless DNA sequences that it cannot be attributed to anything that resembles ID. If the DNA of a human being or any other organism resembled a carefully constructed computer program, with neatly arranged and logically structured modules, each written to fulfill a specific function, the evidence of ID would be overwhelming. In fact, the genome resembles nothing so much as a hodgepodge of borrowed, copied, mutated, and discarded sequences and commands that have been cobbled together by millions of years of trial and error against the relentless test of survival. It works, and it works brilliantly, not because of intelligent design but because of the great blind power of natural selection to innovate, to test, and to discard what fails in favor of what succeeds.

As for Dembski's explanatory filter, because it assumes probabilities that cannot be determined in practice, this is nothing more than a thought experiment, not something that can be used practically in science. Furthermore, rejecting all regulatory hypotheses and chance events (in steps one and two) assumes that we know them all, which, of course, we do not. But even if we did, and rejected them all, the design inference does not follow. Design, as it is commonly defined even by IDers, means purposeful and intelligent creation, not simply the elimination of regularity and chance. In other words, design is not simply a default conclusion when all else fails to explain. Design requires positive evidence, not simply the rejection of negative evidence. Finally, even if positive evidence for design were presented, by the logic of the explanatory filter, it is reasonable to apply the filter to the design claim. Assuming regularity and chance are rejected for the design claim, the logical conclusion would be that the design was designed, and that design was designed, ad infinitum, including and especially the Intelligent Designer himself!

9. Science Education and Debating Evolution. *Creation theory, especially ID theory, provides an alternative to evolution theory, and science education involves hearing both sides of a debate. Science textbooks should carry disclaimers, such as this one now posted inside every biology textbook used in Georgia public schools: "This textbook contains material on evolution. Evolution is a theory, not a fact, regarding the origin of living things. This material should be approached with an open mind, studied carefully and critically considered."*

As innocuous as this argument sounds, it is easily seen through, even by evangelical Christians such as former U.S. president and Georgia resident Jimmy Carter: "As a Christian, a trained engineer and scientist, and a professor at Emory University, I am embarrassed by Superintendent Kathy Cox's attempt to censor and distort the education of Georgia's students. The existing and long-standing use of the word 'evolution' in our state's textbooks has not adversely affected Georgians' belief in the omnipotence of God as creator of the universe. There can be no incompatibility between Christian faith and proven facts concerning geology, biology, and astronomy. There is no need to teach that stars can fall out of the sky and land on a flat Earth in order to defend our religious faith."

There is another, deeper flaw in this argument from "debate," and that is, which creation theory is to be debated with evolution theory? The world's foremost expert on creationism, Eugenie Scott, executive director of the National Center for Science Education, explains, "I encourage people to reject the creation/evolution dichotomy and recognize the creation/evolution continuum. It is clear that creationism comes in many forms. If a student tells a teacher, 'I'm a creationist,' the teacher needs to ask, 'What kind?'" To get our minds around this concept, Scott has developed a powerful visual heuristic in the creation/evolution continuum (for a fuller explication go to http://www.natcensied.org/) that reveals at least ten different positions on this continuum, including:

Flat Earthers. The shape of the earth is flat because a literal reading of the Bible demands it. The earth is shaped like a coin, not a ball. Scientific views are of secondary importance.

Geocentrists. They accept that the earth is spherical, but deny that the sun is the center of the solar system. Like flat earthers, they reject virtually all of modern physics and chemistry as well as biology.

Young-Earth Creationism (YEC). Few classical YECs interpret the flat-earth and geocentric passages of the Bible literally, but they reject modern physics, chemistry, and geology concerning the age of the earth, and they deny biological descent with modification. In their view, the earth is from six to ten thousand years old.

Old Earth Creationism (OEC). From the mid-1700s on, the theology of Special Creationism has been harmonized with scientific data and theory showing that the earth is ancient. Theologically, the most critical element of Special Creationism is God's personal involvement in creation; precise details of how God created are considered secondary.

Gap Creationism. One of the better-known accommodations of religion to science was Gap or Restitution Creationism, which claimed that there was a large temporal gap between Genesis I:1 and I:2. Articulated from about the late eighteenth century on, Gap Creationism assumes a pre-Adamic creation that was destroyed before Genesis I:2, when God re-created the world in six days, and created Adam and Eve. A time gap between two separate creations allows for an accommodation of the proof of the ancient age of the earth with Special Creationism.

Day-Age Creationism. This model accommodates science and religion by rendering each of the six days of creation as long periods of time—even thousands or millions of years instead of merely twenty-four hours long. Many literalists have found comfort in what they think is a rough parallel between organic evolution and Genesis, in which plants appear before animals, and human beings appear last.

Progressive Creationism (PC). The PC view blends Special Creationism with a fair amount of modern science. Progressive Creationists such as Dr. Hugh Ross, of Reasons to Believe ministries, have no problems with scientific data concerning the age of the earth, or the long period of time it has taken for the earth to come to its current form. PCs generally believe that God created "kinds" of animals sequentially; the fossil record is thus an accurate representation of history because different animals and plants appeared at different times rather than having been created all at once.

Intelligent Design Creationists (IDC). Intelligent Design Creationism is a lineal descendant of William Paley's argument from design, which asserted that God's existence could be proved by examining his works. The finding of order, purpose, and design in the world is proof of an omniscient designer.

Evolutionary Creationism (EC). God the Creator uses evolution to bring about the universe according to his plan. From a scientific point of view, evolutionary creationism is hardly distinguishable from Theistic Evolution, which follows it on the continuum. The differences between EC and Theistic Evolution lie not in science but in theology, with EC being held by more conservative (evangelical) Christians.

Theistic Evolution (TE). God creates through evolution. Astronomical, geological, and biological evolution are acceptable to TEs. They vary in whether and how much God is allowed to intervene. Other TEs see God as intervening at critical intervals during the history of life (especially in the origin of humans), and they in turn come closer to PCs. In one form or another, TE is the view of creation taught at mainline Protestant seminaries, and it is the official position of the Catholic Church.

The continuum diffuses the debate and forces the uninitiated into thinking through which of the many positions most appeals to them based on their religious beliefs. With so many mutually exclusive creationist doctrines all claiming infallibility and final truth, a logical default position to fall to is science because it never makes such absolutist truth claims. In science, all conclusions are provisional, subject to new evidence and better arguments, the very antithesis of religious faith.

10. Nonreligious Commitments. *According to William Dembski, mathematician, philosopher, theologian, and author of* Intelligent Design, No Free Lunch, Design Inference, *and other works that form the canon of the New New Creationism: "Scientific creationism has prior religious commitments whereas intelligent design has not."*

Baloney. In the same book Dembski also wrote: "Christ is never an addendum to a scientific theory but always a completion." This is what I call the farce of ID. The primary reason we are experiencing this peculiarly American phenomenon of *evolution denial* (the doppelgänger of *Holocaust denial*) is that a small but vocal minority of religious fundamentalists misread the theory of evolution as a challenge to their deeply held religious convictions. Make no mistake about it. Creationists do not want equal time. They want all the time. Theirs is a war on evolution in particular and science in general, and they are as fanatical in their zeal as any religious movement of the past millennium. Listen to the voice of Phillip Johnson, the fountainhead of the modern ID movement, at a February 6, 2000, meeting of the National Religious Broadcasters in Anaheim, California: "Christians in the twentieth century have been playing defense. They've been fighting a defensive war to defend what they have, to defend as much of it as they can. It never turns the tide. What we're trying to do is something entirely different. We're trying to go into enemy territory, their very center, and blow up the ammunition dump. What is their ammunition dump in this metaphor? It is their version of creation."

Johnson uses another metaphor: a wedge. In his 2000 book *The Wedge of Truth*, he writes: "The Wedge of my title is an informal movement of like-minded thinkers in which I have taken a leading role. Our strategy is to drive the thin end of our Wedge into the cracks in the log of naturalism by bringing long-neglected questions to the surface and introducing them to public debate." This is not just an attack on naturalism, it is a religious war against all of science. "It is time to set out more fully how the Wedge program fits into the specific Christian gospel (as distinguished from generic theism), and how and where questions of biblical authority enter the picture. As Christians develop a more thorough understanding of these questions, they will begin to see more clearly how ordinary people— specifically people who are not scientists or professional scholars—can more effectively engage the secular world on behalf of the gospel."

Finally, in a sermon to the Unification Church Jonathan Wells, author of *Icons of Evolution*, revealed his true motives for studying evolutionary theory: "Father's [the Reverend Sun Myung Moon's] words, my studies, and my prayers convinced me that I should devote my life to destroying Darwinism, just as many of my fellow Unificationists had already devoted their lives to destroying Marxism. When Father chose me (along with about a dozen other seminary graduates) to enter a Ph.D. program in 1978, I welcomed the opportunity to prepare myself for battle."

Let me be blunt (as if I could be even more). It is not coincidental that ID supporters are almost all Christians. It is inevitable. *ID arguments are reasons to believe if you already believe.* If you do not, the ID arguments are untenable. But I would go further. If you believe in God, you believe for personal and emotional reasons, not out of logical deductions. IDers, like the creationists of old, are not only Christians, they are mostly male and educated. In an extensive study on why people believe in God, Frank Sulloway and I discovered that the number one reason people give for their belief is the good design of the world. When asked why they think *other* people believe in God, however, the number one reason offered was emotional need and comfort, with the good design of the world dropping to sixth place. Furthermore, we found that educated men who already believed in God were far more likely to give rational reasons for their belief than were women and uneducated believers. One explanation for these results is that although, in general, education leads to a decrease in religious faith, for those people who are educated and still believe in God there appears to be a need to justify their beliefs with rational arguments.

What is really going on in the ID movement is that highly educated religious men are justifying their faith with sophisticated scientistic arguments. This is old-time religion dressed up in newfangled language. The words change but the arguments remain the same. As Karl Marx once noted: "Hegel remarks somewhere that all great, world-historical facts and personages occur, as it were, twice. He has forgotten to add: the first time as tragedy, the second as farce." The creationism of William Jennings Bryan and the Scopes trial was a tragedy. The creationism of the ID theorists is a farce.

History's Heretics

Who and What Mattered in the Past?

IN 1975 MADAME TUSSAUD'S Waxwork Museum in London took a poll of 3,500 international visitors who came through its doors, asking them to rank the five most hated and the five most beloved people in history. The results were as follows:

Most Hated	*Most Loved*
1. Adolf Hitler	1. Winston Churchill
2. Idi Amin	2. John F. Kennedy
3. Count Dracula	3. Joan of Arc
4. Richard Nixon	4. Robin Hood
5. Jack the Ripper	5. Napoleon

It is an interesting list because it speaks as much for current popularity as it does historical impact (no doubt today Osama bin Laden would make the most hated list). Such a lineup, however meaningless it may be historically, brings to mind a fascination we have with comparisons and lists. Just who were the most important people and events in history?

Comparisons can be objectionable and such best-and-worst lists that demand comparisons onerous. In a culture that esteems greatness and exalts eminence, however, while at the same time sharing pluralistic judgment values, comparisons frequently turn into arguments, and lists morph into endlessly debated rankings. Such lists are ubiquitous. The *Book of Lists* and the *Guinness Book of Records* are both popular top-selling publications. In the final six months of 1999 millennial fever spiraled upward in a headlong rush into naming the most important people and events of the century and millennium. Most were embarrassingly ahistorical and celebrity-centric, with famous folks from the 1980s and 1990s dominating a listing of what was supposed to be the most significant names of the last thousand years.

On the more serious side, *Time* magazine ran a five-part series on "the most influential people of the century" that included extensive essays on each of their chosen hundred. In the "Heroes and Icons" edition, Howard Chua-Eoan opined, "We need our heroes to give meaning to time. Human existence, in the words of T. S. Eliot, is made up of 'undisciplined squads of emotion,' and to articulate our 'general mess of imprecision of feeling' we turn to heroes and icons—the nearly sacred modules of humanity with which we parse and model our lives." Yet he noted the dilemma of discerning true heroes from mere icons: "Iconoclasm is inherent in every icon, and heroes can wear different faces in the afterlives granted them by history and remembrance." Therein lies the problem with any list that includes persons for whom history has yet to grant proper perspective, such as anyone named from the past couple of decades.

Time senior editor Richard Stengel identified another problem inherent in such lists at the close of their "100" series: "We all know about Carlyle's Great Man theory of history, but what about the Creepy Guy Behind the Curtain theory of history or the Meddlesome Housemaid Who Spikes the Punch theory or the Wife Who Whispers in the Great Man's Ear theory? History is written by the victors, but what of those who called in sick that day? Or those who opted not to play? What of the individual who performed one small act that set in motion a great, grand tumult of actions that changed history?" Stengel's pick for the creepiest guy of the twentieth century is Gavrilo Princip, who triggered the outbreak of the Great War when he assassinated Archduke Francis Ferdinand and his wife in Sarajevo in 1914. Lee Harvey Oswald comes to mind as well, yet neither made anyone's list. In the end (literally their final issue of the century), *Time* named Albert Einstein as their "Person of the Century," and I can only say hallelujah to that.

Life magazine, in a special double issue in the fall of 1997, wisely focused on events rather than people as the factor that most changed the world over the past millennium. (The editors got it right in parsing the time frame from A.D. 1001–2000, but they nevertheless could not wait until the end of 2000, for marketing departments often trump editorial departments in such sales-driven businesses.) Two dozen editors consulted scores of experts and ranked the top hundred by a number of criteria, including how many people an event affected and how daily life changed after the event. And an odd bunch it is with some obvious picks (Gregorian calendar, gunpowder) and others, well, strangely quixotic (*Don Quixote* as the first modern novel, Haitian independence).

100. The Gregorian calendar: 1582
 99. Rock and Roll: 1954
 98. Deciphering the Rosetta stone: 1799
 97. Modern Olympics: 1896
 96. First modern novel (*Don Quixote de la Mancha*): 1605
 95. First public museum (Ashmolean, Oxford): 1683
 94. Defeat of the Spanish Armada: 1588
 93. Anesthesia: 1846
 92. Rise of the Ottoman Empire: 1453
 91. Haiti independence: 1804
 50. Mechanical clock: 1656
 10. Compass for open ocean navigation: 1117
  9. Hitler comes to power: 1933
  8. Declaration of Independence: 1776
  7. Gunpowder weapons: c. 1100
  6. Germ theory of disease: 1882
  5. Galileo discovers moons of Jupiter: 1610
  4. Industrial revolution: 1769
  3. Protestant reformation: 1517
  2. Columbus makes first contact with New World: 1492
  1. Gutenberg and the printing press: 1454

For some reason probably influenced by our humanity, we prefer talking (usually gossiping) about people more than events, and the *Life* editors could not resist the temptation to weigh in on *who* mattered, as well as what. They concluded their millennium celebration with the following ranking (surprisingly, given the lack of attention typically paid to science, five of the top ten are scientists, seven if you count Jefferson and Leonardo):

100. Carolus Linnaeus
99. Kwame Nkrumah
98. Ibn-Khaldun
97. Catherine de Médicis
96. Jacques Cousteau
95. Santiago Ramón y Cajal
94. John von Neumann
93. Leo Tolstoy
92. Roger Bannister
91. Nelson Mandela
50. Dante Alighieri
10. Thomas Jefferson
9. Charles Darwin
8. Louis Pasteur
7. Ferdinand Magellan
6. Isaac Newton
5. Leonardo da Vinci
4. Galileo Galilei
3. Martin Luther
2. Christopher Columbus
1. Thomas Edison

Fame and celebrity may land you on someone's arbitrary list, but is it history's list of those who *really* made a difference? Consider CBS News's book *People of the Century: The One Hundred Men and Women Who Shaped the Last One Hundred Years*, published in a coffee-table format. The list starts off reasonably enough, with Freud, Roosevelt, Ford, the Wright brothers, Gandhi, Lenin, Churchill, Einstein, Sanger, and Fleming filling out the early decades. But among the most important people of the later decades were, the editors conclude, Jim Henson, the Beatles,

Pelé, Bruce Lee, Oprah Winfrey, Princess Diana, Bob Dylan, Muhammad Ali, Steven Spielberg, and (one hopes for nothing more than a touch of humor) Bart Simpson. Famedom infiltrates even the august halls of CBS News.

Finally, weighing in on 1999's obsession with history's lists is the *New York Times Magazine*, whose editors presented six themes (in six special Sunday issues), including:

1. "The Best of the Millennium" (fifty-eight writers offer their picks of the most significant events of the past thousand years)
2. "Women: The Shadow Story of the Millennium" (those who have for too long been left off history's lists)
3. "Into the Unknown" (the great adventures of the last thousand years)
4. "New Eyes" (eminent living artists reinterpret history)
5. "The Me Millennium" ("from anonymity then to radical selfishness now")
6. "Time Capsule" (what we should save for people in 3000)

In the final issue, Jared Diamond, author of *Guns, Germs, and Steel*—a history of the past thirteen thousand years—suggested a clever neologism, *apertology* ("the science of opening"): "Just imagine an apertologist in the year 1000 trying to predict who would end up opening a capsule in 2000. That would have been a no-brainer: the Chinese, of course!" Based on who and what mattered in 1000, today's world should be dominated by the Chinese. But as Diamond shows, predicting history's future lists is even harder than constructing lists about what already happened.

Such lists, while of no import to the work of professional historians, do contain some value if, for no other reason, they stimulate us to think about who and what really matters. Besides, it is fun to wrangle with the list maker's choices. For example, what are we to make of the following list from Ashley Montagu, anthropologist, author, and social commentator, of his "ten worst well-known human beings in history"?

1. Attila the Hun
2. Hitler
3. Kaiser Wilhelm II

4. Ivan the Terrible
5. Idi Amin
6. Heinrich Himmler
7. Stalin
8. Caligula
9. Nixon
10. Comte J. A. de Gobineau (theory of the superior Aryan race)

In a different twist on the list game, the *Herald Tribune* picked the top ten news stories of the twentieth century.

1. Kennedy's assassination
2. Bolshevik revolution
3. Moon landing
4. Hiroshima atomic explosion
5. Hitler's launching of WW II
6. Wall Street crash
7. Birth control pill
8. Pearl Harbor
9. Independence of India
10. Lindbergh's flight

Jumping the gun in 1984 (inspired by Mr. Orwell?), the popular science magazine *Science 84* offered the most important "discoveries that shaped our lives" in the twentieth century. Their criterion was that the discovery (actually most are inventions) had to be within the fields of science and technology, and had to have a "significant impact on the way we live or the way we think about ourselves and our world." They were:

1. Plastics and nylon
2. IQ test
3. Einstein's relativity theory
4. Blood typing and transfusion
5. Pearson's chi-square statistics
6. Vacuum tube
7. Controlled breeding of crops
8. Powered flight

9. Penicillin
10. Discovery of ancient man in Africa
11. Atomic fission/bomb
12. The big bang theory
13. DDT
14. Television
15. The Pill
16. Computer
17. Psychoactive drugs
18. Transistor
19. DNA
20. Laser

Enumerateology: The Science of Making Lists

It turns out that such listing tendencies—let's call it *enumerateology*, or the science of making lists—are not new nor are they constrained to our century. History's lists themselves have a history (see figures 12.1– 12.4). In the sixteenth century, for example, the Flemish artist Johannes Stradanus considered these nine discoveries the most outstanding of that age:

1. The New World
2. The magnetic compass
3. Gunpowder
4. Printing press
5. Mechanical clocks
6. Guaiac wood (mistakenly thought to cure syphilis)
7. Distillation
8. Silk cultivation
9. Stirrups

Stradanus began a tradition that would have a long lineage extending to our time (although we seem to prefer round numbers). Lists broad in scope and deep in time include rankings of the most influential people in all of history, not just within a particular age or century. Michael Hart's

Figure 12.1.   The printing press descends from heaven. The tool that gave rise to the other tools of learning—the printing press—is shown symbolically descending from heaven. The press is carried by Minerva and Mercury who will present it first to Gutenberg, master printer in Germany, who will then pass it along to master printers in Holland, Italy, England, and France.

Figure 12.2.   Nine discoveries that changed the world, in the sixteenth century. Nine discoveries and inventions considered the most significant of his time, by the Flemish artist Johannes Stradanus (1523–1605): 1. the New World; 2. the magnetic compass; 3. gunpowder; 4. the printing press; 5. mechanical clocks; 6. Guaiac wood (mistakenly thought to cure syphilis); 7. distillation; 8. silk cultivation 9. stirrups.

Figure 12.3.   Académie Royale des Sciences' selection of the most important works of the age. A 1698 composite perspective of the scientific and technological work in progress at the Académie Royale des Sciences, in Paris, France. Mathematics, physics, and astronomy are heavily emphasized.

Figure 12.4.   Frontispiece from Francis Bacon's 1620 *Instauratio Magna*. This classic engraving from Francis Bacon's 1620 "Great Restoration" of knowledge through the new instrument of science is symbolic of what mattered in history—knowledge and the courage to use it to sail from the known into the unknown. The ships represent the tools of scientific knowledge that carry the explorers (scientists) past the Pillars of Hercules (literally, the Straits of Gibraltar), separating the known (the Mediterranean) from the unknown (the Atlantic).

The 100, for instance, dares to present an "importance ranking" of all time from top ten:

1. Muhammad
2. Newton
3. Christ
4. Buddha
5. Confucius
6. Saint Paul
7. Ts'ai Lun
8. Gutenberg
9. Columbus
10. Einstein

To the bottom ten:

91. Peter the Great
92. Mencius
93. Dalton
94. Homer
95. Elizabeth I
96. Justinian
97. Kepler
98. Picasso
99. Mahavira
100. Bohr

In response to Hart's list of mostly white males, Columbus Salley wrote *The Black 100*, which includes an importance ranking from the top ten:

1. Martin Luther King
2. Frederick Douglass
3. Booker T. Washington
4. W. E. B. DuBois
5. Charles H. Houston
6. Richard Allen and Absalom Jones
7. Prince Hall

8. Samuel Cornish and John Russwurm
9. David Walker
10. Nat Turner

To the bottom ten:

91. Ruby Dee and Ossie Davis
92. Harry Belafonte
93. Marian Wright Edelman
94. Marian Anderson
95. Colin Powell
96. Doug Wilder
97. Ron Brown
98. Clarence Thomas
99. Black Power
100. Rosa Parks

The *Wall Street Journal* in 1982 published the results of a ranking of the ten most important developments in all history, as selected by a group of 350 research and development executives. This group came up with:

1. Wheel
2. Bow and arrow
3. Telegraph
4. Electric light
5. Plow
6. Steam engine
7. Vaccine
8. Telephone
9. Paper
10. Flush toilet

Lists, rankings, and comparisons of this nature are a pastime for many thinkers. Even renowned scholars have occasionally dabbled in such trivialities. Hannah Arendt, for example, in *The Human Condition*, opens a chapter with the following rather bold statement: "Three great events stand at the threshold of the modern age and determine its character: the discovery of America and the ensuing exploration of the whole earth; the Reformation; the telescope and the development of a new science that considers the nature of the earth from the viewpoint of the universe."

Mortimer J. Adler made his reputation, in part, as editor of the Great Books of the Western World, a fifty-four-volume set (read list) of the greatest authors in history. Furthermore, included with this set is the Syntopicon, a listing of the 102 (no more, no less) greatest ideas in history, and what each of the great authors had to say about any or all of them. From this project Adler has spun off several other "list" books, including the *Six Great Ideas* (Truth, Goodness, Beauty, Liberty, Equality, Justice).

Similarly, Jacob Bronowski listed the significant developments in the "ascent of man," in his book of the same title. His thirteen chapters are really a chronological directory of the thirteen great ideas in human cultural evolution. Bronowski's well-thought-out docket, summarized from each chapter, included:

1. Domestication of plants and animals
2. Jericho, the wheel, the horse
3. Architecture and the arch
4. Widespread use of the metals copper, bronze, and iron
5. Mathematics
6. Telescope, heliocentrism, scientific method
7. Theories of gravity and relativity
8. Steam engine/industrial revolution
9. Theory of evolution
10. The table of chemical elements/atomic structure
11. Gaussian statistics, Heisenberg uncertainty principle
12. Mendelian genetics, DNA
13. Awareness of our own identity and ignorance

In similar fashion Richard Hardison, in *Upon the Shoulders of Giants*, noted that "the great builders of the world can be divided into two classes, those who build with stone and mortar and those who build with ideas." Hardison's roster of the fourteen builders and ideas that have "shaped the Modern Mind" included:

1. Copernicus and heliocentricism/Galileo and the scientific method
2. Darwin and evolution
3. Wundt and determinism applied to the study of man
4. Newton and gravity
5. Pascal and probabilities/statistics
6. Calvin, Luther, Protestant Reformation, the rise of humanism and capitalism
7. Pasteur/germ theory, vaccination, pasteurization, consequences of overpopulation
8. Rousseau/democracy
9. Einstein/relativity
10. Gutenberg/the printing press
11. Ford/assembly-line production
12. Faraday/the electric generator, Maxwell/electromagnetic field equations, Edison/the lightbulb
13. Babbage/the computer
14. Awareness of the challenge to our own identity and ignorance

The well-known futurist and author of the best-selling book *Future Shock* Alvin Toffler followed this work with another sweeping survey of our society in *The Third Wave*. This book includes a brief and broad history of culture that Toffler sees as moving in three massive waves. Toffler invokes the metaphor of the "wave" in the sense that Frederick Jackson Turner did in *The Frontier in American History* and Norbert Elias did in *The Civilizing Process*. A wave is "advancing integration over several centuries" (Elias), or a movement or migration of people (Turner) such as in the settlement of the American West—the pioneers, then the farmers, then the "third wave," the business interests. In Toffler's words, "Once I began thinking in terms of waves of change, colliding and overlapping, causing conflict and tension . . . it changed my perception of change itself." In Toffler's analysis, the three waves in human history are: (1) the Agricultural Revolution; (2) the Industrial Revolution; and (3) the Technological (electronic) Revolution. We are currently experiencing the third wave and, according to Toffler, these three waves are in collision in various parts of the world (especially in the so-called developing nations), causing much strife and conflict.

A number of thinkers have made history's lists their vocational modus operandi. James Burke, for example, through the medium of television, has made a career of challenging viewers and readers to consider the forces in history that have shaped the modern world. In his first work in this genre, *Connections*, Burke drew on the various links between ideas, inventions, discoveries, people, and "forces that have caused change in the past, looking in particular at eight recent innovations which may be most influential in structuring our own futures." Burke's eight were:

1. The atomic bomb
2. The telephone
3. The computer
4. The production-line system of manufacture
5. The aircraft
6. Plastics
7. The guided rocket (which can carry atomic bombs)
8. Television

Burke summons the allegory of the "tools" of history that act as the triggers of change: "Each one of these is part of a family of similar devices, and is the result of a sequence of closely connected events extending from the ancient world until the present day. Each has enormous potential for man's benefit—or his destruction."

Burke takes a "great event" approach to history, and in his second work, *The Day the Universe Changed*, he presents the eight most significant "days," or moments of change in Western history, to reveal how an alternate view triggered by an invention or discovery changed the lives of everyone in the culture. These are not just scientific revolutions. Burke describes ideological revolutions that affect all aspects of life. In his own words (my enumeration), Burke writes that: "Each chapter begins at the point where the view is about to shift."

1. In the eleventh century before the extraordinary discoveries by the Spanish Crusaders
2. In the Florentine economic boom of the fourteenth century before a new way of [perspective] painting [that provided the ability to project and predict geometric space that] took Columbus to America
3. In the strange memory-world that existed before printing changed the meaning of "fact"

4. With sixteenth-century gunnery developments that triggered the birth of modern science
5. In the early eighteenth century when hot English summers brought the Industrial Revolution
6. At the battlefield surgery stations of the French revolutionary armies where people first became statistics
7. With the nineteenth-century discovery of dinosaur fossils that led to the theory of evolution
8. With the electrical experiments of the 1820s, which heralded the end of scientific certainty

Will and Ariel Durant succeeded in popularizing history through their monumental eleven-volume *The Story of Civilization*. In four decades of work the Durants compiled over ten thousand pages, covering the great ages of humanity from *Oriental Heritage* to *The Age of Napoleon*. After ten thousand pages covering ten thousand years, Durant was asked who he thought history's ten greatest thinkers were. He answered:

1. Confucius
2. Plato
3. Aristotle
4. Thomas Aquinas
5. Copernicus
6. Francis Bacon
7. Newton
8. Voltaire
9. Kant
10. Darwin

The celebrated science fiction author and science popularizer extraordinaire Isaac Asimov offered his version of the ten most important scientists in history. Alphabetically sorted (to avoid the controversy and near impossibility of ranking, no doubt) they were:

1. Archimedes
2. Darwin
3. Einstein
4. Faraday
5. Galileo
6. Lavoisier
7. Maxwell
8. Newton
9. Pasteur
10. Rutherford

In 1990, when he was managing editor of the *New York Times*, Clifton Daniel organized a project entitled *Chronicle of the 20th Century* that condensed the history of the century into the ten most important news headlines from his paper. In chronological ranking they are:

1. Man's First Flight in a Heavier-Than-Air Machine (December 17, 1903)
2. The Great Powers Go to War in Europe (August 1, 1914)
3. The Bolshevik Revolution in Russia (November 7, 1917)
4. Lindbergh Flies the Atlantic Alone (May 21, 1927)
5. Hitler Becomes Chancellor of Germany (January 30, 1933)
6. Roosevelt Is Inaugurated as President (March 4, 1933)
7. Scientists Split the Atom, Releasing Incredible Power (January 28, 1939)
8. The Nightmare Again—War in Europe (September 1, 1939)
9. Surprise Japanese Bombing of Pearl Harbor (December 7, 1941)
10. Men Land on Moon (July 20, 1969)

In the modern world we have a fascination with the great events of the past. It is stimulating to take a *vista grande* backward in time and pick out, from the billions of people, events, and chance occurrences, those that really made a difference. What we would not give in an instant of fleeting fancy to travel back to a day of critical circumstance and experience that moment with all its historical importance, as seen with twenty-twenty hindsight.

R&D magazine recognizes significant technological developments each year, and on their twentieth anniversary they let their readers select the top technological developments of all time. Each reader was given three votes—first, second, and third—out of nineteen advances they had previously voted as the most significant. Their ranking of the nineteen from top to bottom, including percentage of votes, was:

1. Harnessing electricity, 18.5%
2. Antibiotics, 13%
3. Computer, 10.7%
4. Vaccines, 9.9%
5. Internal combustion engine, 7.4%
6. Genetic engineering, 5.9%
7. Solid-state technology, 5.6%

8. Transistor, 4.1%
9. Quantum mechanics, 4%
10. Nuclear power, 3.6%
11. Special theory of relativity, 2.9%
12. Nuclear weapons, 2.6%
13. Superconductivity, 2.5%
14. Television, 2.3%
15. Telephone, 2.2%
16. Birth control pill 2.1%
17. Laser, 1.2%
18. Radio diode, 0.9%
19. Electron microscopy, 0.6%

Clearly electricity dominates the list. Taking all electrically related inventions together the total is 37.3 percent. Biological-related developments, however, were not far behind at 30.9 percent. Quantum mechanics and special relativity were the only two theories on the list, though the theories behind the other developments, while not mentioned, are of equal or greater importance.

Marshall McLuhan, the famed communications expert who wrote extensively on the influence of the media on modern history ("the medium is the message"), made a presentation of a type of history's list in the form of the "ten most potent extensions of man." They included:

1. Fire	6. The sword
2. Clothing	7. Print
3. The wheel	8. Electric telegraph
4. The lever	9. Electric light
5. Phonetic alphabet	10. Radio/TV

McLuhan asserts that the alphabet is an "extension of language," and radio and television are "extensions of the central nervous system." He lists the telegraph as the predecessor to the telephone, and notes that "print makes everyone a reader" and "Xerox makes everyone a publisher."

From the sublime to the ridiculous, the *Book of Lists* gives us William Manchester's "ten favorite dinner guests from all history." It is an absurd historical list, but it is interesting because Manchester is an independent scholar and historian, and has therefore not been restricted to working within the bounds of certain historical periods or places. His interests are broad and eclectic.

1. Newton
2. Elizabeth Tudor
3. Freud
4. Emma Hamilton
5. George Bernard Shaw

6. Oscar Wilde
7. Napoleon
8. Jane Austen
9. Goethe
10. H. L. Mencken

History as Accumulation

Top one hundred or top ten lists, by definition, are limiting. The advantage they do offer, however, is that they can give us a broad and sweeping glance at the richly detailed and highly convoluted connections of history. History's lists are to historical data what science's theories are to scientific facts. They allow us to search for general principles, trends, highlights, and conglomerations, within a myriad of seemingly disparate bits of information. To paraphrase a quote from the physicist Poincaré, a group of historical facts is no more a history than a pile of bricks is a building. History's lists are the blueprints of the past.

Any list of scientific and technological advances in history is an "interactive" list, in that the later scientists and technologists had the earlier thinkers and cultures from which they could benefit. The steps that are early on the list make possible the later ones. Though they are listed individually, taken as a whole the steps later on the list are "richer" than the earlier selections, in the sense Will Durant described in *The Lessons of History*:

> The heritage that we can now more fully transmit is richer than ever before. It is richer than that of Pericles, for it includes all the Greek flowering that followed him; richer than Leonardo's, for it includes him and the Italian Renaissance; richer than Voltaire's, for it embraces all the French Enlightenment and its ecumenical dissemination. If progress is real . . . it is not because we are born any healthier, better, or wiser than infants were in the past, but because we are born to a richer heritage, born on a higher level of that

pedestal which the accumulation of knowledge and art raises as the ground and support of our being.

Knowledge is cumulative in this interactive sense. Ideas feed upon one another and are connected in innumerable ways. Any listing, whether it is the top ten, hundred, five hundred, thousand, will necessarily leave out a step in the development of some idea. But the selection of one invention over another, for example, does not really eliminate the unchosen step. Ideas evolve, not unlike species, through descent with modification. Later ideas are different from, but still connected by ancestry to, earlier ideas. Past thinkers and civilizations do not really die; they evolve into another form, sometimes more advanced, sometimes less. Durant observes:

> Greek civilization is not really dead; only its frame is gone and its habitat has changed and spread; it survives in the memory of the race, and in such abundance that no one life, however full and long, could absorb it all. Homer has more readers now than ever in his own day and land. The Greek poets and philosophers are in every library and college; at this moment Plato is being studied by a hundred thousand discoverers of the "dear delight" of philosophy overspreading life with understanding thought.

The Hundred: A Personal Ranking

Inspired by this survey of history's lists, what follows is a personal ranking, chronologically, of the hundred, the twelve, and the single most influential event in the history of science and technology. They were selected for their historical significance in light of their impact on the modern Western world. These are the discoveries, inventions, books, and events within the disparate fields of science and technology that have shaped the modern mind and created today's complex and dynamic world. The basic criterion in sorting through the many thousands of choices was the number of people influenced. The more impact the selection had on society, whether direct or indirect, the greater the likelihood it made the list. Although by nature I am unabashedly inclined to favor science and technology over all other human endeavors as the most significant, I think it is clear from this extensive survey of history's lists that scientific discoveries and technological advances have by far and away done more to shape history. History's heretics may come from many walks of life, but heretical scientists and technologists have influenced our past, present, and future more than anyone else.

In this sequential listing the obvious temptation to list all hundred in order of importance has been avoided. While in some cases it would be easy to justify and defend ranking one invention over another (e.g., the lightbulb over the astrolabe), it would be an impossible task to rank the entire hundred and maintain any sense of objectivity or credibility. The joy in lists and comparisons is quibbling with the author. For example, how could Michael Hart possibly rank Kepler, whom historians of science call the "father of modern astronomy," at number 97, while he places John F. Kennedy at number 80 or Simón Bolívar at number 46? But this critique simply reflects my bias for science over politics. For their impact value on our culture, it would be hard to argue against any of the *Herald Tribune*'s top news stories of the twentieth century being on a list of some sort. But Kennedy's assassination ranked six places ahead of the birth control pill?! How could Durant leave out Augustine, or Asimov exclude Aristotle? The reader will probably ask the same questions of my list below.

The entries on my list are broken down into the following major categories within the framework of science and technology: Discoveries, Inventions, Books, and Events. Examples include:

Discoveries: evolution, relativity, heliocentricity, non-Euclidian geometry, circulation, anaesthesia, etc.

Inventions: pendulum clock, the Pill, laser, telephone, lightbulb, gunpowder, compass, etc.

Books: *The Origin of Species, Principia, Elements of Geology, Almagest, Elements of Geometry,* etc.

Events: moon landing, founding of natural Greek philosophy, powered flight, development of scientific method, etc.

The list is chronological, beginning in 10,000 B.C.E., and excludes prehistoric and early human "inventions" such as language, fire, the bow and arrow, stone tools, clothing, etc. The top hundred are listed first, followed by the top twelve, and finally my selection of the single most important contribution to science and technology. No scholar or scientist in his right mind would undertake such a precarious enterprise of collapsing ten thousand years, a hundred billion people, and a thousand trillion events into a list of one hundred. We proceed.

1. Domestication of animals and plants (ends hunter-gatherer lifestyle): dog: c. 10,000; goat/sheep: 8,700; pig: 7,500; cattle: 6,800; chicken: 5,000; horse: 4,400; mule/donkey/camel: 3,000; cat: 2,500; wheat/barley: 8,600; potatoes: 8,000; rice: 7,000; sugarcane: 7,000; squash: 6,000; maize: 5,700; grapes: 4,000; olives: 3,500; cotton: 3,300
2. Widespread use of metals: copper c. 4,000; bronze c. 3,000; iron c. 2,000
3. Plow in common use c. 3500–3000
4. Wheel in common use c. 3500–3000
5. Writing (cuneiform) c. 3300 (Sumeria)
6. Founding of natural Greek philosophy c. 600–500 (Ionian Miletus)
7. Geometry (Pythagoras) 581–497 (helped found Western science)
8. Hippocratic collection/oath 460–377 (sixty medical books)
9. Atomic theory ("Atomists") c. 400 (Democritus)
10. Aristotelian logic and naturalism c. 335
11. Elements of geometry (Euclid)
12. Zero used in mathematics c. 300 (Babylon)
13. Alexandrian library burned 47 (burned again in C.E. 391)
14. Waterwheel in productive use 14 (Roman)
15. *Guide to geography* and *Almagest* c. 90–160 (Ptolemy—geocentrism; grid lines on map bend with curvature of earth; "Terra Incognita" triggers later explorations)
16. Arch c. 100 (Roman architecture)
17. Lateen sail (Arab) c. 100–200 (improves speed/maneuverability of ships)
18. Paper (China) 105 (Ts'ai Lun)
19. Descriptive anatomy/experimental physiology 129–199 (Galen—four humors identified)
20. Stirrup (Europe) c. 700–800 (changed knight warfare)
21. Astrolabe 850 (Arab—navigation)
22. Magnetic compass in common use 1125 (China and Europe)
23. Flying buttress c. 1200 (European architecture)

24. Coal discovered 1233 (England—source of fuel)
25. Decimal notation (Arab) 1250 (replaced Roman numerals)
26. *De Computo Naturali* 1264 (Roger Bacon—experimental "natural" science)
27. Marco Polo's explorations 1271–1295 (to China and the Far East)
28. Verge-and-Foliot clock 1280 (improves timing, mechanizes society)
29. Gunpowder c. mid-thirteenth century (earlier in China)
30. Spectacles 1303 (unknown inventor)
31. Perspective geometry c. 1400 (Toscanelli/Brunelleschi—cartography and art; world map with grid lines/perspective used by Columbus)
32. Printing press/movable type 1454 (Gutenberg—Germany)
33. Portable/inexpensive book 1495 (Aldus Manutius—Italy, "pocket" book)
34. Circumnavigation of the earth 1522 (Magellan's expedition)
35. Organic theory of disease 1527 (Paracelsus)
36. *On the Revolution of the Celestial Spheres* 1543 (Copernicus)
37. Structure of the human body 1543 (Vesalius)
38. Telescope 1600 (Lippershey/Holland; 1608—Galileo's astronomical observations)
39. Laws of planetary motion 1618–1619 (Kepler)
40. Patent laws 1623 (England)
41. Circulatory system 1628 (Harvey)
42. Development of scientific method 1629 (Francis Bacon's *Novum Organum* initiates development, followed by contributions from Galileo, Descartes, Newton)
43. Dialogue concerning two chief world systems 1632 (Galileo supports heliocentricism)
44. *Discourse on Method* and *Geometrie* 1637 (Descartes, mechanistic/reductionistic philosophy of science; analytic geometry)
45. Coke (England) 1640 (improved fuel from coal)
46. Probability theory 1654 (Pascal)
47. Pendulum clock 1657 (Huygens vastly improves timing devices)
48. Microscope 1665 (unknown inventor; Hooke's *Micrographia*)
49. Balance spring timepiece 1674 (Hooke/Huygens—accurate time keeping/calculation of longitude)

50. Differential/integral calculus 1675 (Newton/Leibniz)
51. Mathematical principles of natural philosophy 1687 (Newton's *Principia*)
52. Crucible steel process 1740 (Huntsman)
53. Steam engine 1775 (Watt, Newcomen, Papin)
54. Elements of chemistry 1789 (Lavoisier)
55. Cotton gin 1793 (Whitney)
56. Vaccination 1798 (Jenner)
57. Essay on population 1798 (Malthus identifies overpopulation problem)
58. Standardization of machine parts 1800 (Brunel/Maudslay)
59. Normal distribution curve 1801 (Gaussian statistics)
60. Jacquard loom 1801 (programmed cards later used in computers)
61. Food preservation 1809 (Appert)
62. Principles of geology 1830–1833 (Lyell)
63. Electric generator/motor 1831 (Faraday's Law)
64. Computer 1833 (Babbage—analytical engine)
65. Telegraph 1837 (Morse/Wheatstone/Cooke)
66. Surgical anesthesia 1846 (Morton—ether)
67. Electromagnetic field equations 1855 (Maxwell)
68. *The Origin of Species* 1859 (Darwin)
69. Germ theory 1861 (Pasteur)
70. Antiseptic surgery 1865 (Lister—carbolic acid)
71. Laws of heredity 1865 (Mendel)
72. Periodic table of elements 1869 (Mendeleyev)
73. Internal combustion engine/auto 1876 (Otto)
74. Telephone 1876 (Bell)
75. Invention factory 1876 (Edison)
76. Incandescent light globe 1879 (Edison)
77. Radioactivity 1896 (Becquerel)
78. Radio 1896 (Marconi)
79. Quantum theory proposed 1900 (Planck's theory of the atom)
80. Powered flight 1903 (Wright brothers)
81. Special/general theory of relativity 1905/1915 (Einstein)
82. Vacuum tube 1907 (De Forest)
83. Assembly-line mass production 1913 (Ford)

84. Origin of continents and oceans 1915 (Wegener's continental drift)
85. Iconoscope 1924 (Zworykin—television)
86. Bosch process 1925 (Carl Bosch)
87. Liquid rocket fuel 1926 (Goddard)
88. Uncertainty principle 1927 (Heisenberg)
89. Penicillin 1928 (Fleming/Florey/Chain)
90. Cyclotron 1931 (Lawrence—particle accelerator)
91. Nylon patented 1937 (Carothers/DuPont Co.—plastics, synthetics, etc.)
92. Jet engine 1937 (Whittle)
93. Electrophotography 1938 (Carlson/Xerox—photocopy)
94. Nuclear detonation 1945 (Los Alamos)
95. Transistor 1947 (Shockley/Brattain/Bardeen)
96. Laser 1951 (Townes)
97. Genetic code 1953 (Crick/Watson/Wilkins)
98. The Pill 1960 (Pincus)
99. Manned moon landing 1969 (*Apollo* 11)
100. Genetic engineering/cloning 1973 (Köhler/Milstein)

The Top Twelve

1. Founding of natural Greek philosophy/Aristotelian logic c. 600–300 B.C.E.
2. Printing press/movable type 1454 (Gutenberg)
3. *On the Revolution of the Celestial Spheres* 1543 (Copernicus)
4. Development of scientific method 1629 (Francis Bacon's *Novum Organum* initiates development—Galileo, Descartes, Newton)
5. Mathematical principles of natural philosophy 1687 (Newton's *Principia*)
6. Steam engine 1775 (Watt)
7. Electric generator 1831 (Faraday)
8. Computer 1833 (Babbage's analytical engine)
9. *The Origin of Species* 1859 (Darwin—theory of evolution by natural selection)
10. Germ theory 1861 (Pasteur)

11. Manned/powered flight 1903 (Wright brothers)
12. Special/general theory of relativity 1905/1915 (Einstein)

The Single Most Important Contribution in History

1. *The Origin of Species* 1859 (Darwin—theory of evolution by natural selection)

PART IV

SCIENCE
AND THE CULT OF
VISIONARIES

13

The Hero on the Edge of Forever

Gene Roddenberry, Star Trek,

and the Heroic in History

HISTORIANS AND BIOGRAPHERS have explained the origin of the heroic in two dramatically different ways. At one end of the spectrum heroes are "great men"—seminal thinkers, brilliant inventors, creative authors. At the other end, heroes are historical artifacts of their culture—ordinary people thrust into positions of power and fame that might just as well have gone to others. The first archetype is represented by Thomas Carlyle in his *Heroes, Hero-Worship, and the Heroic in History*: "Universal History, the history of what man has accomplished in this world, is at bottom the history of the great men who have worked here. Worship of a hero is transcendent admiration of a great man." The second archetype is seen in such Marxist writers as Friedrich Engels: "That a certain particular man, and no other, emerges at a definite time in a given country is naturally a pure chance, but even if we eliminate him there is always a need for a substitute, and the substitute . . . is sure to be found."

Although such polarities are held by relatively few, the central claims of both contain an element of truth. History's heroes may be great individuals, but all individuals, great or not, are—indeed must be—culturally bound; where else could they act out the drama of their heroics?

In this sense, history and biography may be modeled as a massively contingent multitude of linkages across space and time, where the hero is molded into, and helps to mold the shape of, those contingencies. For Sidney Hook, in his classic study *The Hero in History*, a hero is "the individual to whom we can justifiably attribute preponderant influence in determining an issue or event whose consequences would have been profoundly different if he had not acted as he did." History is not strictly determined by the forces of the weather or geography, demographic trends or economic shifts, class struggles or military alliances. The hero has a role in this historical model of interacting forces—between unplanned contingencies and forceful necessities.

Contingencies are the sometimes small, apparently insignificant, and usually unexpected events of life—for want of a horseshoe nail the kingdom was lost. Necessities are the large and powerful laws of nature, forces of economics, trends of history—for want of 100,000 horseshoe nails the kingdom was lost. Elsewhere I have presented a formal model describing the interaction of these historical variables (see chapter 10), summarized in these brief definitions: *contingency* is taken to mean *a conjuncture of events occurring without perceptible design*; *necessity* is *constraining circumstances compelling a certain course of action*. Leaving either contingencies or necessities out of the biographical formula, however, is misleading. History is composed of both; therefore it is useful to combine the two into one term that expresses this interrelationship—*contingent-necessity*—taken to mean *a conjuncture of events compelling a certain course of action by constraining prior conditions*.

Contingent-necessity is actually an old concept in new clothing. The Roman historian Tacitus was uncertain "whether it is fate and unchangeable necessity or chance which governs the revolutions of human affairs," where we have "the capacity of choosing our life," but "the choice once made, there is a fixed sequence of events." Karl Marx offered this brilliantly succinct one-liner (from *The Eighteenth Brumaire*): "Men make their own history, but they do not make it just as they please; they do not make it under circumstances chosen by themselves, but under circumstances directly found, given and transmitted from the past."

A question arises from this: Can we find a repeatable pattern in historical sequences that demonstrates when and where contingencies and necessities will dominate in the life of an individual? Contingency and necessity vary in both influence and sequential position within any given historical sequence according to what is called the model of contingent-necessity, which states: *In the development of any historical sequence the role of contingencies in the construction of necessities is accentuated in the early stages and attenuated in the later.*

To tell the story of someone's life, then, the biographer must understand not only the contingencies and the necessities of the individual's history, as well as the history around him, but the nature and timing of their interaction. The biographer must see which conjunctures of events compelled certain courses of action (contingencies constructing necessities), and which constraining prior conditions determined future actions (necessities creating outcomes). In other words, we must know not only the life and the times of the person, but how this life and those times interacted, when, and in what sequence to produce the outcome in question. As an exemplar of both biography and this particular philosophy of biography I suggest the life of Gene Roddenberry told by his biographer David Alexander, who has skillfully reconstructed the contingencies and necessities—the life and times—of the creator of *Star Trek*. And, as I hope to show, Gene Roddenberry has done the same through his fictional starship voyages in his humanistic vision of humanity's future.

Reviewing episodes of the original series of *Star Trek* today, one is struck by the almost absurdly simplistic design of set and costumes (made worse by the cartoonish colors of the sixties-era clothing). Yet the show resonated across generations to become one of the most remarkable television phenomena in the medium's history. The reason is relatively simple: Roddenberry was an enlightened storyteller who baldly addressed the deepest issues in science, religion, philosophy, politics, and current events. He was adept at placating television executives and advertisers, while simultaneously maintaining a commitment to promoting science and humanism. David Alexander was one of Gene Roddenberry's closest friends and was handpicked by Roddenberry to be his authorized biographer. One might thus expect a dearth of dirt. But no, it is here, warts and all, although to be fair, additional warts have been added by others since the original publication of Alexander's biography in 1993. The entertainment journalist Joel Engel, for example, has painted a far less flattering portrait of Roddenberry in his 1994 biography, *Gene Roddenberry: The Myth and the Man Behind* Star Trek. For me, anyway, little of this dirt matters in the overall impact of the man and his humanistic philosophy. Knowing that he sometimes took credit that was arguably due others, or that he had an extramarital affair, or that his disputes with writers and network executives were not always handled with perfect professional aplomb only adds to the humanness of this humanist. So he was morally obtuse on occasion—I hold up Roddenberry and his literary corpus not as a moral beacon toward which we should strive, but for what we can learn from his creative visage of our future.

The contingencies of Roddenberry's life unfolded within the necessities of his times, starting with the Second World War, in which Roddenberry served as a bomber pilot flying unprotected B-17 sorties in the South Pacific. After the war he found employment as a commercial pilot for Pan Am and was involved in the single worst flight disaster in Pan Am's history. Remarkably, Roddenberry survived the horrific crash, and as a consequence he was flooded with thoughts on death, God, and the meaning of existence. Roddenberry's two marriages, his ambitious but naive start in television, his struggle to launch *Star Trek* and then keep it on the air after the second season, his battles with studios over the films and novels based on his series, and many more anecdotes show that *Star Trek* could never have been so successful if it had not been produced by a man whose life experiences gave him an insight into human nature appreciated by the millions who tuned in to get more than just science fantasy. In short, the contingencies that constructed the necessities of Roddenberry's life show conclusively: no Gene Roddenberry, no *Star Trek*.

Roddenberry was, first and foremost, a humanist who used science to communicate his deeper intentions. Raised a Baptist, in his early teens he grew skeptical of certain theological claims, such as the Eucharist, or Holy Communion. "I was around fourteen and emerging as a personality," he told David Alexander for an interview in *The Humanist*. "I had never really paid much attention to the sermon before. I listened to the sermon, and I remember complete astonishment because what they were talking about were things that were just crazy. It was Communion time, where you eat this wafer and are supposed to be eating the body of Christ and drinking his blood. My first impression was, 'This is a bunch of cannibals they've put me down among.'" The scales fell from his now skeptical eyes. God became "the guy who knows you masturbate," Jesus was no different from Santa Claus, and religion was "nonsense," "magic," and "superstition." For Roddenberry, science holds the key to the future, and as such he strove for accuracy. In the following letter, for example, Roddenberry responds to the legendary science writer Isaac Asimov, who had taken him to task for a scientifically inaccurate statement in one of the early episodes. We see Roddenberry properly respectful of this literary giant, yet appropriately defensive of his fictional child.

> In the specific comment you made about *Star Trek*, the mysterious cloud being "one-half light-year outside the Galaxy," I agree certainly that this was stated badly, but on the other hand, it got past a Rand Corporation physicist who is hired by us to review all of our stories and scripts, and further, got past Kellum deForest Research who is also hired to do the same job.
>
> And, needless to say, it got past me.
>
> We do spend several hundred dollars a week to guarantee scientific accuracy. And several hundred more dollars a week to guarantee other forms of accuracy, logical progressions, etc. Before going into production we made up a "Writer's Guide" covering many of these things and we send out new pages, amendments, lists of terminology, excerpts of science articles, etc., to our writers continually. And to our directors. And specific science information to our actors depending on the job they portray. For example, we are presently accumulating a file on space medicine for De Forest Kelley who plays the ship's surgeon aboard the USS *Enterprise*. William Shatner, playing Captain James Kirk, and Leonard Nimoy, playing Mr. Spock, spend much of their free time reading articles, clippings, SF stories, and other material we send them.
>
> Despite all of this we do make mistakes and will probably continue to make them. The reason—Thursday has an annoying way of coming up once a week, and five working days an episode is a crushing burden, an impossible one. The wonder of it is not that we make mistakes, but that we are able to turn out once a week science fiction which is (if we are to believe SF writers and fans who are writing us in increasing numbers) the first true SF series ever made on television.

(Nevertheless, scientific errors crept into the show such that today books and Web sites have been devoted to *Star Trek* bloopers.)

Later, after Roddenberry's fame skyrocketed along with the series (now in syndication with multiple spin-off series and feature films, generating hundreds of millions of dollars), the *Star Trek* creator was asked to comment on any number of deep issues, especially those of a religious and spiritual nature. Roddenberry was a humanist in the purest sense of the word—he had a deep love of humanity and held out the greatest hope for our future, without depending on a higher power to achieve happiness. He offered these thoughts on God and UFOs to Terrance Sweeney, for a book entitled *God &*, published in 1985:

> I presume you want an introspective look at this. I think I've gone through quite an ordinary series of steps in life. I began as most children began, with God and Santa Claus and the tooth fairy and the Easter Bunny all being about the same thing. Then I went through the things that I think sensitive people go through, wrestling with the thoughts of Jesus—did he shit? Did he screw? I began to dare to believe that God wasn't some white beard. I began to look upon the miseries of the human race and to think God was not as simple as my mother said. As nearly as I can concentrate on the question today, I believe I am God; certainly you are, I think we intelligent beings on this planet are all a piece of God, are becoming God. In some sort of cyclical non-time thing we have to become God, so that we can end up creating ourselves, so that we can be in the first place.
>
> I'm one of those people who insists on hard facts. I won't believe in a flying saucer until one lands out here or someone gives me photographs. But I am almost as sure about this as if I did have facts, although the only test I have is my own consciousness.

Roddenberry's philosophical humanism was tempered with a practical realism necessary in dealing with television studios. MGM and CBS, for example, initially turned *Star Trek* down. NBC bought the show but had to pay for another pilot because the first one—"The Cage"—was "too cerebral." They were also worried about Spock's pointed ears, airbrushed out of the publicity stills to appease conservative network executives and potential advertisers. Thus was created a second pilot (another first in television history)—"Where No Man Has Gone Before"—which became a series classic. Finally, the first *Star Trek* episode aired on the night of my twelfth birthday, Thursday, September 8, 1966. The program was "The Man Trap," ironically the sixth episode filmed and not nearly the same quality as the two pilots. "The Cage" was later made into a two-part episode (the only one of the original series) and included black-and-white footage and some of the original actors who never appeared in the regular series.

One of the best known and arguably the finest program of the seventy-nine episodes in the original series—"The City on the Edge of Forever"—was written by the highly acclaimed science fiction writer Harlan Ellison. In 1994, *Entertainment Weekly* ranked all seventy-nine episodes of the original *Star Trek* series—"The City on the Edge of Forever" was number 1. The next year, *TV Guide* published their "100 Most Memorable Moments in TV History," for which "The City on the Edge of Forever" ranked number 68. The televised show won a Hugo Award, and Ellison's original script—similar in its core but considerably different in details from what aired on television—won the Writers Guild of America Award for the Most Outstanding Teleplay for 1967–68. For $200 you can even purchase a pewter and porcelain desktop sculptured model of the famed time portal from the episode, produced by the Franklin Mint.

The history of "The City on the Edge of Forever" has become a point of gossipy controversy in *Trek* lore, so it bears brief synopsis. Most *Star Trek* histories and memoirs touch on the controversy, and most get the story wrong. To set the record straight, Harlan Ellison published an entire book on the subject in 1996, entitled (so you don't miss it) *Harlan Ellison's The City on the Edge of Forever: The Original Teleplay that Became the Classic Star Trek Episode*. The book is backed by copious documentation in support of his version of this tiny slice of television minutia, along with an unstoppable logorrhea made tolerable by Ellison's inimitable style. To wit, on Roddenberry's tendency to take credit for other people's work, Ellison writes:

> The supreme, overwhelming egocentricity of Gene Roddenberry, that could not permit him to admit anyone else in his mad-god universe was capable of grandeur, of expertise, of rectitude. And his hordes of Trekkie believers, and his pig-snout associates who knew whence that river of gold flowed . . . they protected and buttressed him. For thirty years.
>
> If you read all of this book, I have the faint and joyless hope that at last, after all this time, you will understand why I could not love that aired version, why I treasure the Writers Guild award for the original version as that year's best episodic-dramatic teleplay, why I despise the mendacious fuckers who have twisted the story and retold it to the glory of someone who didn't deserve it, at the expense of a writer who worked his ass off to create something original, and why it was necessary—after thirty years—to expend almost 30,000 words in self-serving justification of being the only

person on the face of the Earth who won't let Gene Roddenberry rest in peace.

To quote Captain Kirk from *Star Trek II: The Wrath of Khan*: "Don't mince words . . . what do you *really* think?"

Ellison submitted an initial treatment of the episode on March 21, 1966, and a revised treatment on May 13, 1966. The original script, touching on issues in the forefront in the 1960s, involved an *Enterprise* crew member dealing "infamous and illegal Jillkan dream-narcotics, the Jewels of Sound," used (and abused) to handle the stress and boredom of long-term space flight. In addition, Ellison had a legless World War I veteran assist Kirk and Spock in their historical mission. Finally, Ellison developed the character of Captain Kirk as more morally complex than he had been in early episodes, in this case freezing at the crucial moment that demanded action. Roddenberry rejected these elements, apparently believing that drugs, cripples, and moral obtuseness were not good images for noble crews and heroic captains. To be fair, when Ellison penned his treatment there were only two scripts from which to work—"The Cage" and "Where No Man Has Gone Before"—and the show was undergoing rapid and dramatic changes in its early stages. Therefore, says Alexander, Roddenberry rewrote it himself after other staff writers failed to bring it up to the show's standards.

Ellison disputes Alexander's take on the matter, presenting a complex story that involves a number of writers exerting their influence—least of all Roddenberry. "I rewrote the script, I rewrote it again, I worked on it at home and on a packing crate in Bill Theiss's wardrobe room in Building 'E' and when Gene kept insisting on more and more changes, and when I saw the script being dumbed up, I couldn't take much more." The script was redacted by Steve Carabatsos, Gene Coon, and Dorothy C. Fontana, and "fiddled" (says Ellison) by Roddenberry. In the end, Ellison was sufficiently displeased with the final product and requested that his name be substituted with his Writer's Guild Association registered pseudonym, Cordwainer Bird, "which everyone in the industry knew was Ellison standing behind this crippled thing saying *it ain't my work* and sort of giving the Bird to those who had mucked up the words," Harlan explained in the third person, then confessing "but Gene called me and made it clear he'd blackball me in the industry if I tried to humiliate him like that; and I went for the okeydoke. I let my name stay on it." To this day Ellison's name stands on this most famous of *Star Trek* episodes, which originally aired on NBC on April 6, 1967.

I have now read "The City on the Edge of Forever" in all of its editions. Although I must admit that Harlan's original script is richer, more complex, and more morally compelling than what aired on television, childhood memories create powerful preferences. To a thirteen-year-old boy, William Shatner cuts a heroic jib while Joan Collins fulfills youthful fantasies. Regardless, the story is not just another science fiction tale. Three decades later that episode came to represent to me a deeper theme of historical change. Since historians cannot go back in time to alter a jot or tittle here and there and observe the subsequently changed outcome, such historical fantasizing must be left to those who dwell in fiction. While science has yet to devise a way to travel backward in time, science fiction has no trouble at all (usually by encountering an "anomaly" in the space-time continuum), and the theme of removing an individual from the historical picture to trigger a different result has become one of the mainstays of the genre.

The discussion of fictional stories in this context may be defended as contributing to our scientific understanding of the nature of causality in history and biography. There is a role for thought experiments in all sciences—to broaden our perspective and deepen our understanding of a subject through the consideration of novel possibilities (see chapter 10). The Austrian physicist and philosopher of science Ernst Mach considered thought experiments a form of empiricism because they derive from past experiences creatively rearranged. He notes that physicists' thought experiments such as rolling balls down frictionless planes serve a useful purpose in establishing principles that can be tested experimentally in other ways. Einstein is the best-known example of a scientist many of whose most significant ideas were worked out in thought long before they were tested experimentally. He called these his "gedanken experiments." But he was only doing what so many scientists do, and not just physicists. Economists are famous for their ceteris paribus assumption of "all other things being equal" thought experiments about the economy, conditions that rarely exist in the real world but serve a useful purpose for manipulating variables in theory that could not be rearranged in reality.

How do we conduct a historical thought experiment? Mach explains: "When experimenting in thought it is permissible to modify unimportant circumstances in order to bring out new features in a given case." He also warns: "But it is not to be antecedently assumed that the universe is without influence on the phenomenon in question." Mach's admonition is well taken, and "The City on the Edge of Forever" presents a healthy balance between specific circumstances (contingencies) and universal effects (necessities), and how their interaction shapes altered histories. The story line offers a simple but powerful message about the actions of individuals and the contingencies of life, while recognizing that these occur within the context of influencing larger forces.

In "The City on the Edge of Forever" the theme of the present and future contingently linked to the past—where every individual and event has some relative effect (usually small, occasionally large)—is played out in this science fiction thought experiment. On Stardate 3134.0, an *Enterprise* landing crew is beamed down to a planet to rescue their temporarily psychotic physician, gone mad because of a cordrazine drug–induced accident (in Ellison's original script drug abuse led to an onboard murder, for which the convicted crewman was condemned to spend the rest of his life on this barren planet). While on the planet, the landing crew investigates an anomalous source of energy causing "ripples" in the fabric of time. The crew comes upon an arch-shaped rock structure that Spock's tricorder readings indicate is on the order of ten thousand centuries (one million years) old. Surging with power, it appears to be the single source of the time ripples. Kirk's rhetorical inquiry of "What is it?" produces a response: "I am the guardian of forever. I am my own beginning, my own ending." Spock, in his archetypical omniscience explains: "A time portal, Captain. A gateway to other times and dimensions."

Acknowledging Spock's correct analysis, the "guardian" gives them a centuries-long glimpse into history, "a gateway to your own past, if you wish." What they witness is every historian's fantasy—a replay of history with the opportunity to go back and relive the past. "Strangely compelling, isn't it?" Kirk queries. "To step through there and lose oneself in another world." But as they ponder the imponderable, the still drug-crazed physician, Dr. McCoy, appears from behind a rock and leaps through the time portal, which promptly terminates its historical replay. "Where is he?" Kirk asks. The guardian replies: "He has passed into . . . what was." A landing-party member who was in contact with the ship at the moment McCoy jumped through the time portal suddenly loses all communications. "Your vessel, your beginning, all that you knew is gone," explains the guardian. McCoy, in the manner of the increasingly popular "what if" genre of historical thought experiments, in something akin to the matricide problem of time travel, altered the past in a manner that erased the *Enterprise* from history. (Why the landing crew would remain is not explained, nor is the matricide paradox resolved, for if you go back in time and kill your mother before you were born, then you erase yourself from the continuum that allowed you to alter that original past. If McCoy changed the past in such a way that the *Enterprise* and her crew no longer existed, then he could not have jumped through the time portal to erase that history.)

Matricide paradox aside (which it must be for time travel scenarios to exist in science fiction), the solution is that Kirk and Spock must return to the past just before McCoy jumped through the portal and prevent him from doing whatever it was he did to alter the future. Exactly when they should jump through the portal and where on Earth McCoy landed is the problem. Spock, the science officer, offers a solution: "There could be some logic to the belief that time is fluid, like a river: currents, eddies, backwash." Kirk concurs: "And the same currents that swept McCoy to a certain time and place might sweep us there too."

Together they leap through the portal and arrive in New York City circa the 1930s and take refuge in the basement of a mission run by the angelic Edith Keeler (Joan Collins through soft-filter lenses, whose character Ellison patterned after "Sister" Aimee Semple McPherson, the first evangelist to employ the power of radio in the 1930s to reach the masses). Spock constructs a crude computer out of vacuum tubes in order to replay his recording of the guardian's passage of time (which he was producing with his tricorder when McCoy jumped) to see if they can determine what McCoy did to alter the future. In the process, Spock gets a brief glimpse of the immediate future in a newspaper headline of that year that read "Social Worker Killed," beneath which appears a picture of Edith Keeler. The image fades as a power surge burns out the tubes. Kirk arrives, after a plot-thickening encounter with Keeler in which the two appear to be falling in love.

"I may have found our focal point in time," Spock explains to him. "Captain, you may find this a bit distressing." A replay of the now-repaired computer, however, reveals a different newspaper caption six years into the future: "F.D.R. Confers with Slum Area 'Angel.'" Above the headline is a photo of Edith Keeler. "We know her future. Within six years from now she'll become very important, nationally famous," Kirk enthuses. Spock gives him the alternative. "Or, Captain, Edith Keeler will die, this year. I saw her obituary. Some sort of a traffic accident." Kirk is understandably confused. "You must be mistaken. They both can't be true."

Spock explains the paradox that such time-travel thought experiments present. "Captain, Edith Keeler is the focal point in time we've been looking for, the point in time that both we and Dr. McCoy have been drawn to." The now-enlightened Kirk says to no one in particular, "She has two possible futures, and depending on whether she lives or dies all of history will be changed. And McCoy . . ." Spock finishes the sentence: ". . . is the random element." Kirk queries: "In his condition, does he kill her?" Spock answers with a double paradoxical problem: "Or, perhaps he prevents her from being killed. We don't know which. Captain, suppose we discover that in order to set things straight again, Edith Keeler must die?"

Meanwhile, McCoy makes his appearance in New York, near Keeler's mission, hungover from his drug overdose. As he enters the mission to recover, he narrowly misses seeing Kirk and Spock as they descend to the basement where Spock has repaired his crude computer, to glance into the future one last time. "This is how history went after McCoy changed it," Spock explains to Kirk. "Here, in the late 1930s, a growing pacifist movement whose influence delays the United States entry into the Second World War. While peace negotiations dragged on, Germany had time to complete its heavy-water experiments."

In this altered version of history, Germany is the first to develop the atomic bomb. Mounted on V-2 rockets, these bombs allow Germany to defeat the Allies and win the war, altering conditions so that centuries hence the *Enterprise* and her crew never existed. Since it was Edith Keeler who founded the peace movement, somehow McCoy came back and prevented her from dying in a street accident. (Of course, if he did she would have erased McCoy from the future, preventing him from returning to the past to save her life, in which case she would have been killed, allowing McCoy to live and save her life, and so on.) She lives, along with the cascading alternate contingencies of Nazi victory. Much to Kirk's dismay, for history to be righted there is only one future Edith Keeler can have. Spock reinforces the overriding necessity: "Save her—do as your heart tells you to do—and millions will die who did not die before."

The stage is set for the final scene. Kirk and Keeler are walking down the street hand in hand. She makes some mention of McCoy staying at the mission, which is the first signal to Kirk that the doctor has arrived. Kirk exhorts Keeler to stay put while he bolts across the street toward the mission in front of which Spock is standing. As he reaches the sidewalk McCoy exits the building and the three reunite in mutual delight of recognition amid this temporal nightmare. A curious Keeler begins her journey into destiny by stepping off the curb and crossing the street, oblivious to a truck approaching at high speed. Kirk turns and sees the immediate danger in time to leap out and save Keeler. Spock insists Kirk stop himself. He does, but McCoy, unaware of the impending consequences, starts toward the street to save the doomed social worker. Kirk grabs him, preventing his intervention. Keeler is killed. In shock, McCoy exclaims: "You deliberately stopped me, Jim. I could have saved her. Do you know what you just did?" Heartbroken and speechless, Kirk allows Spock the final line. "He knows, doctor. He knows."

History was repaired, contingencies righted, necessities in place. Kirk and Spock suddenly appear out of the portal to the surprised landing crew who experienced almost no lapse of time. The *Enterprise* awaits their safe return, and the guardian concludes: "Time has resumed its shape. All is as it was before."

In Ellison's original script, the World War I crippled veteran named Trooper leads Kirk and Spock to a different character named Beckwith (neither appeared in the final televised version) who is the random element in the time line who shows up to save Keeler. Kirk moves to stop him, but then hesitates. "He cannot sacrifice her, even for the safety of the universe," Ellison writes. "But at that moment Spock, who has been out of sight, but nearby, fearing just such an eventuality, steps forward and freezes Beckwith in midstep. Edith keeps going and we QUICK CUT to Kirk as we HEAR the SOUND of a TRUCK SCREECHING TO A HALT. As Kirk's face crumbles, we know what has happened. Destiny has resumed its normal course, the past has been set straight." In an insightful epilogue that would have elevated the televised episode into literary enlightenment, Ellison allows the heroes to reflect on what has just unfolded. "We look at our race, this parade of men and women, and the unbelievable harm and cruelty they do," Kirk opines. "And we sigh, and we say, 'Perhaps our time is past, let the sharks or the cockroaches take over.'

And then, without knowing why, without even thinking of it, the worst among us does the great thing, the noble deed, that spark of impossible human godliness." Spock continues the thought. "Evil can come from Good, and Good from Evil. But the little man . . . Trooper . . ." Kirk: "He was negligible. He fought at Verdun, and he was negligible. And she . . ." Spock finishes the sentence: "No, she was not negligible." "But . . . I loved her . . ." Kirk anguishes. "No woman was ever loved as much, Jim. Because no woman was ever offered the universe for love."

In his introductory remarks on "The City," Harlan disparages the "hero-ification" of such characters as Roddenberry, citing the historian James W. Loewen (from *Lies My Teacher Told Me*): "Heroification . . . much like cal-cification . . . makes people over into heroes. The media turn flesh-and-blood individuals into pious, perfect creatures without conflicts, pain, credibility, or human interest." Indeed, this is a serious problem when our goal is an accurate portrayal of the way things really are. But when the goal is to envision a future of the way things could be, heroification is a virtue. We need heroes. We want heroes.

On January 30, 1993, NASA posthumously awarded Roddenberry the Distinguished Public Service Medal, with a citation that read: "For distin-guished service to the Nation and the human race in presenting the explo-ration of space as an exciting frontier and a hope for the future." Through *Star Trek*, Gene Roddenberry, Harlan Ellison, and others have offered optimism through a heroification of humanity and history. May we all find our own hero on the edge of forever.

14

This View of Science

The History, Science, and
Philosophy of Stephen Jay Gould

IN THE CLOSING DECADES of the twentieth century the genre of popular science writing by professional scientists blossomed as never before, with sales figures to match the astronomical six- and seven-figure advances being sought and secured by literary agents, and paid, however begrudgingly, by major trade publishing houses. Although popular science exposition has a long historical tradition dating at least to Galileo, never has there been such a market for science books, particularly works written for both professional scientists and general audiences interested in the profound implications for society and culture of scientific discoveries.[1] In the 1960s the mathematician Jacob Bronowski's *The Ascent of Man*, based on his popular PBS documentary series of the same name, earned the previously unknown scientist a measure of fame late in his life. In the 1970s the astronomer Robert Jastrow's *God and the Astronomers* landed him in the chair next to Johnny Carson on *The Tonight Show*, but he was soon displaced by astronomer Carl Sagan, who took the genre to new heights when he broke all records for the largest advance ever given for a first-time novel ($2 million for *Contact*). His book *Cosmos*, based on the PBS series watched by half a billion people in sixty nations, stayed on the *New York Times* bestseller list for over a hundred weeks and sold more copies to that date than any English-language science book ever published.[2] So famous did he become that a "Sagan effect" took hold in science, whereby one's popularity and celebrity with the general public was thought to be inversely proportional to the quantity and quality of real science being done.[3] Sagan's biographers have stated unequivocally, based on numerous interviews with insiders, that Harvard's refusal of Sagan's bid for tenure, and the National Academy of Science's rejection of the nomination of Sagan for membership, was a direct result of this "Sagan effect."[4] But even Sagan's popularity and book sales were exceeded in the late 1980s and early 1990s by the mathematical physicist and cosmologist Stephen Hawking, whose book *A Brief History of Time* set new sales standards for science books to come, with a record two hundred weeks on the London *Times*'s hardback bestseller list, and over ten million copies sold worldwide.[5]

Stephen Jay Gould has been a highly successful participant in this salubrious arrangement among scientists, agents, publishers, and readers. With the exception of his first book, which was a monograph on the relationship between development and evolution (*Ontogeny and Phylogeny*), and his last book, which is a technical synthesis of his life's work (*The Structure of Evolutionary Theory*), the twenty books in between were popular science books also written for his colleagues. With this volume of writing have come the corresponding awards and accolades, including a National Magazine Award for his column "This View of Life," several national book awards, dozens of honorary degrees, fellowships, and awards for achievements and service. He has even been called "America's evolutionist laureate."[6] Along with the recognition, of course, has come the requisite criticisms—the tall trees catch the wind—and Gould has had his fill over the years. In 1986, Harvard biologist Bernard Davis accused Gould of "sacrificing scientific integrity to hyperbole for political purposes."[7] The philosopher Daniel Dennett allocated fifty pages of his 1995 book *Darwin's Dangerous Idea* to Gould, calling him "the boy who cried wolf," a "failed revolutionary," and, in uppercase sarcasm, "Refuter of Orthodox Darwinism."[8] Evolutionary biologist Richard Dawkins says punctuated equilibrium is a "tempest in a teapot" and "bad poetic science," and recounts how "after giving lectures in the United States, I have often been puzzled by a certain pattern of questioning from the audience" involving mass extinctions. "It is almost as though the questioner expects me to be surprised, or discomfited, by the fact that evolution is periodically interrupted by catastrophic mass extinctions. I was baffled by this until the truth suddenly hit me. Of course! The questioner, like many people in North America, has learned his evolution from Gould, and I have been billed as one of those 'ultra-Darwinian' gradualists!"[9] Even in the pages of the *New York Review of Books*, a regular venue for Gould's popular writings over the years, the highly regarded evolutionary biologist John Maynard Smith wrote this stinging appraisal:

> Gould occupies a rather curious position, particularly on his side of the Atlantic. Because of the excellence of his essays, he has come to be seen by non-biologists as the preeminent evolutionary theorist. In contrast, the evolutionary biologists with whom I have discussed his work tend to see him as a man whose ideas are so confused as to be hardly worth bothering with, but as one who should not be publicly criticized because he is at least on our side against the creationists. All this would not matter, were it not that he is giving non-biologists a largely false picture of the state of evolutionary biology.[10]

In 1998 evolutionary biologist John Alcock published a no-holds-barred assault on Gould in the flagship journal for evolutionary psychologists, concluding: "I am confident that, in the long run, Gould's polemical essays will be just an odd footnote in the history of evolutionary thought, a history that has been shaped in a wonderfully productive manner by the adaptationist perspective."[11] Likewise, Robert Wright has targeted Gould in such popular publications as *The New Republic*: "A number of evolutionary biologists complain—to each other, or to journalists off the record—that Gould has warped the public perception of their field." In *Slate* he wrote that "Gould is a fraud" and that "among top-flight evolutionary biologists, Gould is considered a pest—not just a lightweight, but an actively muddled man who has warped the public's understanding of Darwinism." And in a *New Yorker* piece Wright called Gould an "accidental creationist" who "is bad for evolution."[12] Finally, the philosopher of science Michael Ruse complained that "it rankles also that Gould does not fight his battles just in the professional journals" and that "it is not just that Gould's ideas are wrong. It is that they are presented as a position of reason and tolerance and common sense, and the outside world believes him. That really irritates."[13]

Indeed, a lot of people seem rankled and irritated by Gould. Are we witnessing another example of the perceived "Sagan effect" generated by

Gould's enormous popularity among general readers and that he is, in fact, a world-class scientist? Or has Gould's influence within science been highly exaggerated by his popular science expositions and, in reality, he will go down as an "odd footnote" in the history of science? I will address these questions in the context of a remark science historian Ronald Numbers once made: "I can't say much about Gould's strengths as a scientist, but for a long time I've regarded him as the second most influential historian of science (next to Thomas Kuhn)."[14] Historians are deeply familiar with Kuhn's work and influence, and most know of the remarkable popularity of Gould's writings on evolutionary theory and related topics. But little attention has been paid to the depth, scope, and importance of Gould's role as historian and philosopher of science, an analysis of which not only illuminates Numbers's striking observation, but helps put Gould's work as a scientist into a larger context. Gould's writings in science, popular science, and history and philosophy of science are tightly interrelated and feed back upon one another as part of a grander strategy Gould has employed. That strategy can best be summed up in a quotation from Charles Darwin, frequently cited by Gould as a sound principle of philosophy: "All observation must be for or against some view if it is to be of any service." Gould has followed Darwin's advice throughout his career and in his extensive writings in both science and history.[15]

The Measure of a Man

The impetus to do a quantitative content analysis of Gould's works began in 1999 while I was researching the "Sagan effect," testing whether Sagan's popular writings really did attenuate his professional output. Data were gathered from Sagan's weighty 265-page curriculum vitae, but such numbers alone, without a context, reveal little. Since he was rejected by the National Academy of Science, I thought it would be instructive to compare Sagan's literary statistics to those of the average NAS member. Unfortunately, no such comparative data are available, so I compared him to several recognized eminent scientists including Jared Diamond, Ernst Mayr, Edward O. Wilson, and Stephen Jay Gould. It turns out that Sagan falls squarely in the middle of this distinguished group in both total career publications (500 versus Gould's 779, Mayr's 714, Diamond's 563, and

Figure 14.1. Stephen Jay Gould

Wilson's 388) and average publications per year (12.5 versus Gould's 22.3, Diamond's 13.4, Mayr's 9.3, and Wilson's 7.6). My data showed that throughout his career, which began in 1957 and ended in December 1996 upon his untimely death, Sagan averaged a scientific peer-reviewed paper per month. The "Sagan effect" is a chimera.

For Gould's career, from his first published paper in 1965 to the end of 2000 when this count was made, with a career total of 902 publications, Gould has seen his name in print at least twice a month for thirty-five consecutive years. Examining the categories in detail, Gould's book total of twenty-two falls one short of Wilson's and nine short of Sagan's (and ties with Mayr), but subdividing the totals by solo versus coauthored and edited/coedited works places Gould far ahead of the others in the solo division at eighteen, compared to Mayr's thirteen, Sagan's twelve, Wilson's nine, and Diamond's five. Figure 14.2 shows Gould's book reviews by

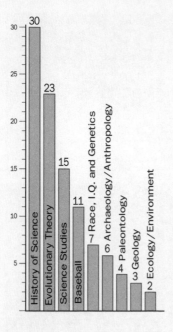

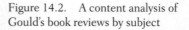

Figure 14.2.   A content analysis of Gould's book reviews by subject

subject, which were classified according to the primary subject of the book under review. Immediately we see that the perception of Gould as scientist and evolutionary theorist is too limiting. In fact, Gould reads and reviews more books on the history of science than any other subject, and adding to that figure the fifteen reviews of books best classified in science studies or the philosophy of science, it makes that figure nearly double that of evolutionary theory. This taxonomy was based on 101 published reviews, many of which contained multiple books under review. Reclassifying this category by books (140) instead of reviews (101) reveals that Gould's favorite subjects are baseball at thirty-five, history of science and evolutionary theory tied at thirty each, and science studies steady at fifteen. The overall conclusion about Gould's professional interests in the history of science and science studies, however, does not change. We shall explore the significance of this interest below.

As for the subject content of Gould's own books, a gross classification scheme puts half in the general category of natural history, with the others divided between history of science/science studies, evolutionary theory, and paleontology/geology.[16] This classification does not tell us much, however, because the books contain too much variation within each one, particularly in the essay collections. To assess his deepest professional interests we must quantify his 479 scientific/scholarly papers, which is presented by maximal taxonomic classification categories (based on the primary subject of each paper) in figure 14.3.

At first glance it would seem that as a scientist and scholar Gould is first and foremost an evolutionary theorist—his 136 papers far outdistancing all other categories. Interestingly, despite the fact that as a scientist Gould is best known for the theory of punctuated equilibrium, he published only fifteen papers on the subject, a mere 3 percent of the total (and, as we shall see, even fewer mentions of the theory are made in his essays). Scanning the graph, however, it becomes clear that a number of these fifteen specialties are obviously allied (for example, paleobiology, paleontology, punctuated equilibrium, paleoanthropology, and geology). By collapsing them into related taxa (figure 14.4) we see that Gould's five primary scientific/scholarly interests are, in order, evolutionary theory, paleontology, history of science, natural history, and interdisciplinary studies.

Classification of Gould's papers was done by considering both the primary subject of the paper and the journal in which it was published. For example, a 1984 paper entitled "The Life and Work of T.J.M. Schopf (1939–1984)," although published in the journal *Paleobiology*, was classified in the history of science because the piece was, first and foremost, an obituary (Schopf was the editor of the 1972 volume on *Models in Paleobiology*, in which Niles Eldredge and Gould first introduced punctuated equilibrium). Similar criteria were used to classify his 1985 article in the journal *Evolution* entitled "Recording Marvels: The Life and Work of George Gaylord Simpson." Later that year, however, Gould published an article in the same journal on "The Consequences of Being Different: Sinistral Coiling in Cerion." Although *Cerion* is the primary subject of Gould's paleontological studies, this paper was classified as evolutionary theory because the main focus was on the evolutionary process of structural change, or allometry, a subject of great interest to Gould and one in which he has published dozens of articles, all classified under evolutionary theory in this taxonomic scheme. By contrast, a 1984 paper on "Covariance Sets and Ordered Geographic Variation in *Cerion* from Aruba, Bonaire and Curaçao: A Way of Studying Nonadaptation," published in *Systematic Zoology*, was classified under paleobiology, as was a 1988 paper on "Prolonged Stability in Local Populations of *Cerion*

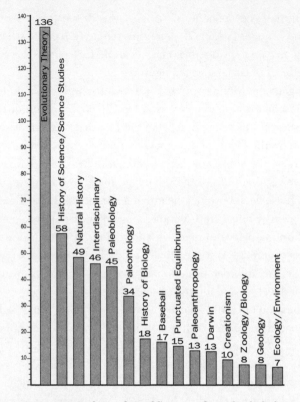

Figure 14.3.   A content analysis of Gould's scientific and scholarly specialties by maximal taxonomic classification

agassizi (Pleistocene-Recent) on Great Bahama Bank," published in the journal *Paleobiology*.

Included in the interdisciplinary category were Gould's writings on baseball and other nonscientific subjects such as writing, teaching, choral singing, and even music (his 1978 "Narration and Précis of J. Dryden, 'King Arthur' for the performance of Purcell's Incidental Music" in particular stands out) based on the seriousness of the scholarship and the amount of original research involved. For example, Gould conducted an extensive analysis of why no one hits .400 in baseball anymore, in which he collected data from baseball archives, computed standard deviations between the worst hitters and the average and the best hitters and the average for over a hundred years of the game's history, and discovered a

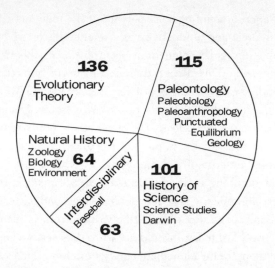

Figure 14.4. A content analysis of Gould's scientific and scholarly specialties by minimal taxonomic classification

statistical trend of improvement in the average play over time such that the best players today, while absolutely as good as, if not better than, players from earlier in the century, are relatively worse compared to today's higher average level of play. This analysis was published in an article entitled "Entropic Homogeneity Isn't Why No One Hits .400 Anymore." Similar reasoning was used to classify a 1979 article entitled "Mickey Mouse Meets Konrad Lorenz," as well as a 1980 article on "Phyletic Size Decrease in Hershey Bars," because they contain light themes with a deep message—long-term evolutionary trends may wash out short-term selective forces, a point that Gould has hammered home time and again in his struggle to balance the adaptationist program with other evolutionary factors.[17] In other words, even Gould's seemingly frivolous writings almost always have a deeper message related to his larger vision of the structure of evolutionary theory and his particular philosophy of science.

Although Gould's insistence that he is "a tradesman, not a polymath"[18] is at least partially supported by the fact he has published 115 scientific papers (24 percent) in his trade field of paleontology and paleobiology, and 136 papers in the allied field of evolutionary theory, clearly Gould is no single-minded fossil digger or armchair theorizer. His 101 papers in the history of science, amounting to 21 percent of the total, not only show his remarkable interest and productivity as a science historian, but also play an integral role in the development of his evolutionary theorizing and science philosophizing. This effect is dramatically borne out in an analysis of his three hundred essays in the popular science magazine *Natural History*. There is no question that "this view of life" is distinctly Gouldian. It is in the essays that we see most clearly the blending of popularization and professionalism. In the prefaces to most of the essay collections, in fact, he makes a spirited defense of the importance of writing to a broader audience without dumbing down. In *Dinosaur in a Haystack*, for example, Gould writes: "I intend my essays for professionals and lay readers alike—an old tradition, by the way, in scientific writing from Galileo to Darwin, though effectively lost today. I would not write these essays any differently if I intended them for my immediate colleagues alone. Thus, while I hope that you will appreciate my respect, our bargain may require a bit more from you than the usual item of American journalism demands."[19] As Gould's consecutive essay streak continued over the decades the demand on general readers' patience and reading skills grew ever greater.

The Streak

Stephen Jay Gould has often stated that his two heroes (other than his father) are Joe DiMaggio and Charles Darwin. Darwin, of course, makes regular appearances in most of Gould's publications, but DiMaggio crops up now and again as well. For a 1984 PBS *Nova* special on Gould, he and his son spent an afternoon playing catch with DiMaggio in a ballpark in the Presidio of San Francisco during which they discussed, of course, Gould's favorite topic of evolutionary trends in life, as well as baseball, including the Yankee Clipper's fifty-six-game hitting streak. A few years later Gould wrote about this "Streak of Streaks," in which he demonstrated through a fairly sophisticated analysis why DiMaggio's streak was so beyond statistical expectation that it should never have happened at all. It was inevitable, then, that Gould's own streak in science writing would be compared favorably to that of Jolt'n Joe's.[20]

Gould's *Natural History* column began in January 1974 with a 1,880-word essay on "Size and Shape," and ended (appropriately, considering

Gould's interest in calendrics and the calculation of the millennium) in the December/January 2000/2001 issue with a 4,750-word essay entitled "I Have Landed."[21] In twenty-seven years Gould wrote approximately 1.25 million words in three hundred essays. The shortest essay was "Darwin's Dilemma" in 1974 at 1,475 words, and the longest (not counting four two-parters, the longest of which was 10,449 words) was "The Piltdown Conspiracy," in 1980 at 9,290 words, for an overall average of 4,166 words. Tracking the length of the essays over time shows that Gould reached his career average by the early 1980s and found his natural length of about 5,000 words by the early 1990s. The late 1990s saw his columns become not only longer (with several six- and seven-thousand-word essays) but more convoluted with multiple layers of complexity.[22]

Much has been made of Gould's literary style, particularly in the essays, which intermingle scientific facts and theory with a large dollop of high- and pop-culture references, foreign language phrases, poetic and literary quotations, and especially biblical passages. Most praise Gould for this linking of science to the humanities, but his critics see something more sinister. John Alcock called it an "ostentatious display of erudition" injected to persuade "many a reader that he is an erudite chap, one whose pronouncements have considerable credibility thanks to his knowledge of foreign languages and connections with Harvard. By advertising his scholarly credentials, Gould gains a debater's advantage, which comes into play when he contrasts his erudition with the supposed absence of same in his opponents." To prove his point Alcock took "a random selection of 20 Gouldian essays" in which he found "nine with at least one word or phrase in German, Latin, or French" and "five of 30 contained quotes from Milton, Dryden, and other literary masters."[23]

Setting aside the insoluble question of how many literary references and foreign language phrases are appropriate, a thorough analysis of all three hundred essays reveals precisely how often Gould utilized these tools in his essays. The foreign phrases total includes Latin (sixteen), French (nine), German (six), and Italian (one). Not included in this count were such commonly used phrases as *natura non facit saltum* ("nature does not make leaps," a phrase used often in nineteenth-century natural history and the subject of an entire essay by Gould), or such everyday expressions as raison d'être. Included were such phrases as *ne plus ultra* ("the ultimate"), *Nosce te ipsum* ("Know thyself"), *Mehr Licht* ("More light"), *Plus ça change, plus c'est la même chose* ("the more things change, the more they remain the same"), and the one Alcock complained about, *Hier stehe ich; ich kann nicht anders; Gott helfe mir; Amen*, Martin Luther's fervent cry of defense for his heresy: "Here I stand; I cannot do otherwise; God help me; Amen." In three hundred essays written over the course of twenty-seven years, a grand total of thirty-two foreign language phrases were employed, amounting to barely 10 percent of the total, or only one in ten essays. If this is a conscious strategy on Gould's part to gain "a debater's advantage," he does not utilize it very often.

Gould's literary references are more frequently employed than foreign phrases at 119 total, with the Bible (fifty-three) outnumbering the next three most quoted of Gilbert and Sullivan (twenty-one), Shakespeare (nineteen), and Alexander Pope (eight) combined. Again, there are no objective criteria on how many literary references are appropriate here, but we can nevertheless discern whether Gould is using them as a strategy to win arguments and wow readers, or if he is trying to make his point through as many avenues available for written prose in an attempt to take science to a broader audience. Not surprising (given Gould's admitted left-leaning upbringing), Karl Marx is often quoted. "Men make their own history, but they do not make just as they please" is used three times, but his favorite is this classic line from the *Eighteenth Brumaire*, quoted no less than seven times: "Hegel remarks somewhere that all great, world-historical facts and personages occur, as it were, twice. He has forgotten to add: the first time as tragedy, the second as farce." The context in which these quotations appear reveal, in fact, that Marx is used by Gould not for show, or for any political or ideological purpose, but directly to bolster his philosophy of science and to reinforce two themata that appear throughout his works—the interaction between contingencies and necessities and the nonrepeatability of historical systems (time's arrow versus time's cycle). "In opening *The Eighteenth Brumaire of Louis Bonaparte*," Gould notes in one essay, "Karl Marx captured this essential property of history as a dynamic balance between the inexorability of forces and the power of individuals." Even Marx's title, Gould explains,

> is, itself, a commentary on the unique and the repetitive in history. The original Napoleon staged his coup d'état against the Directory on November 9–10, 1799, then called the eighteenth day of Bru-

maire, Year VIII, by the revolutionary calendar adopted in 1793 and used until Napoleon crowned himself emperor and returned to the old forms. But Marx's book traces the rise of Louis-Napoleon, nephew of the emperor, from the presidency of France following the revolution of 1848, through his own coup d'état of December 1851, to his crowning as Napoleon III. Marx seeks lessons from repetition, but continually stresses the individuality of each cycle, portraying the second in this case as a mockery of the first.

To drive home the point Gould finishes this thought with a recommendation for scientists to heed the lesson: "This essential tension between the influence of individuals and the power of predictable forces has been well appreciated by historians, but remains foreign to the thoughts and procedures of most scientists."[24]

Similarly, biblical quotations are used to deliver a deeper meaning. In an essay on Charles Doolittle Walcott's misreading of the Burgess Shale fossils and Gould's discussion with paleontologist T. H. Clark (who knew and worked with Walcott) on the "true" meaning of the fossils and on how science works, Gould opines:

> Lives are too rich, too multifaceted for encompassing under any one perspective (thank goodness). I am no relativist in my attitude towards truth; but I am a pluralist in my views on optimal strategies for seeking this most elusive prize. I have been instructed by T. H. Clark and his maximally different vision. There may be no final answer to Pilate's inquiry of Jesus (John 18:37), "What is truth?"— and Jesus did remain silent following the question. But wisdom, which does increase with age, probes from many sides—and she is truly "a tree of life to them that lay hold upon her."[25]

Gould's intellectual pluralism is evident in his literary diversity, and he has chosen many strategies for communicating his answer to Pilate's question.

Essays Thematical

The diversity of Gould's essays was captured poetically by science historian and lyricist Richard Milner in a tribute song (to the music of Gilbert and Sullivan's "My Name Is John Wellington Wells"):

> I write of cladistics
> And baseball statistics
> From dodos and mandrills
> To friezes and spandrels . . .
>
> I write Essays thematical
> Always grammatical
> Asteroids, sesamoids,
> Pestilence tragical
> Ratites, stalactites
> And home runs DiMaggical . . .
>
> I write of Cranial capacity
> Owen's mendacity
> Huxley's audacity
> Galton's urbanity
> FitzRoy's insanity
> How Ernest Haeckel, without an apology
> Faked illustrations about embryology.[26]

Despite the variety, there is a cladistic pattern from which we may discern a literary baüplan. Figure 14.5 presents the results of a complete classification of all three hundred essays into primary, secondary, and tertiary subjects in thirteen different categories.[27]

Starting with the lowest figures we see that Gould almost completely neglects to include his personal hobbies, such as baseball and music, as well as his intellectual child, punctuated equilibrium. He dabbles in ecology and environmental issues, touches on geology and the social and behavioral sciences, and, of course, cannot ignore (but does not dwell on) his own trade of paleontology (and its relations paleobiology and paleoanthropology). Obviously natural history, zoology, and biology are regularly featured, even if only on the secondary or tertiary levels, and since the essay genre is, by definition, personal, Gould does produce a fair amount of social commentary, but predominantly at the tertiary level. What is surprising in this graph is the overwhelming dominance of evolutionary theory and the history of science/science studies, comprising 55 percent of the total. Although the personal nature of essays suggests they need not be

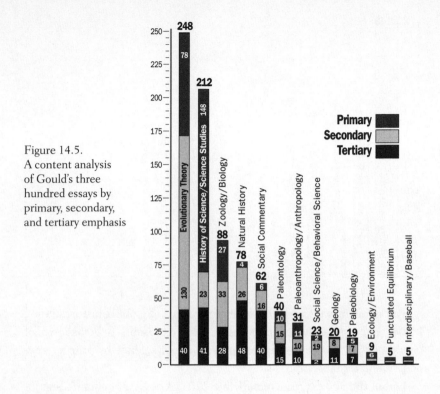

Figure 14.5.
A content analysis
of Gould's three
hundred essays by
primary, secondary,
and tertiary emphasis

taken as seriously as, say, major peer-reviewed journal articles and mono-graphs, clearly Gould is using them to a larger purpose involving not only his interest in theory and history, but as an avenue to generate original contributions to and commentary on both. And it would seem from this graph that Gould is, first and foremost, an evolutionary theorist. Or is he? To explore this question further, figure 14.6 shows the thirteen subject categories collapsed into five, highlighting only the primary subjects.

What is Gould primarily interested in writing about in his essays? While evolutionary theory and the history and philosophy of science once again dominate (comprising 75 percent of the total), they have flip-flopped in dominance from the totals in figure 14.5. That is, the history of science and science studies (which includes philosophy of science) now overwhelm all other subjects, nearly doubling evolutionary theory and totaling almost more than all other categories combined. What is Gould up to when he blends the history and philosophy of science and science studies with evolutionary theory?

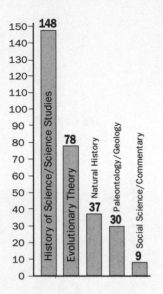

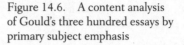

Figure 14.6. A content analysis of Gould's three hundred essays by primary subject emphasis

Part of an answer can be found in an analysis of Gould's historical time frame, and especially in figure 14.7, which presents a breakdown of Gould's essays on the history and philosophy of science by primary, secondary, and tertiary emphasis.[28]

Out of the 300 essays, a remarkable 220 (73 percent) contain a significant historical element, with half (109) in the nineteenth century and nearly a third (64) in the twentieth. Since Gould's primary historical interest is the history of evolutionary theory, we should not be surprised by this ratio since the past two centuries have been the theory's heyday. Yet it is also important to note that the history of evolutionary theory is bracketed in figure 14.7 by the philosophy of science on the right and the relationship between culture and science on the left. All other interests pale by comparison, revealing Gould's intense interest in the interaction of history, theory, philosophy, and culture. For Gould they are inseparable. Doing science also means doing the history and philosophy of science, and as a historian and philosopher of science Gould is intensely interested in the interaction between individual scientists and their culture. This is why there are, in these 220 historical essays, no fewer than seventy-six significant biographical portraits, a number of which include original contributions to the historical record. For example, Gould conducted a thorough analysis of Leonardo's paleontological observations and his theory of the

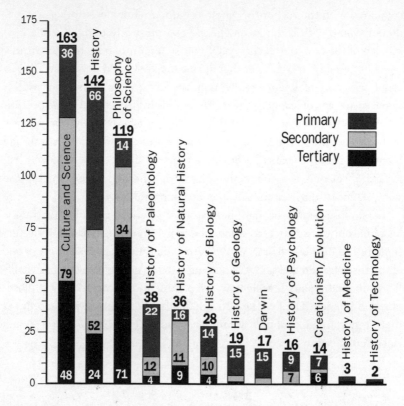

Figure 14.7. A content analysis of Gould's three hundred essays by primary, secondary, and tertiary emphasis in the history and philosophy of science

earth as presented in the Leicester Codex, showing that he was no out-of-time visionary but was instead deeply wedded to the premodern world-view of the sixteenth century.[29]

Gould's work in the history of science can also be seen quantitatively in the annual Current Bibliographies of the History of Science Society journal *Isis*. Although some years are sparse, such as 1991 and 1992 with just three references each and 1997 with only two, other years show Gould outpublishing all other historians with, for example, 24 references in 1986, 16 references in 1988, and 12 in 1989. Gould's overall average reference rate in the *Isis Bibliography* indexes between 1977 and 1999 is 7.34 (169 references in twenty-three years). No other historian comes close to Gould in generating this much history of science, and these figures, conjoined with the rest of this analysis, support Ronald Numbers's equation of Gould with Kuhn as one of the two most influential historians of science of the twentieth century. Of course, quantity does not necessarily equate to quality, and the fact that during his life Gould never developed a cadre of history of science students in the same manner as other professional historians of science may mean that his influence will come posthumously (Gould died on May 19, 2002). To that extent, then, this paper is both prescriptive and descriptive. (Gould reiterated to me his continued frustration over the years that he did not seem to be taken seriously by historians of science, the one community he felt he had not reached to the same extent he did members of science communities.)

Even more important than the history of science in Gould's writings is his philosophy of science, as evidenced in five thematic pairs representing some of the deepest themes in Western thought that appear in every one of the three hundred essays. Classifying Gould's essays into one of five different thematic pairs reveals how inseparable are history, theory, philosophy, and science. The five themata are displayed in figure 14.8, in order of their importance in Gould's writings as shown by the number of essays classified in each:[30]

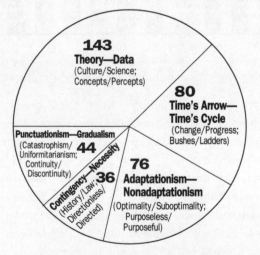

Figure 14.8. Gould's Five Themata.

Frank Sulloway identified the second theme, *Time's Arrow—Time's Cycle*, as an important element in Gould's work: "The more one reviews his writing over the years, the more one sees just how central this and another thematic pair of ideas—continuity and discontinuity—are in his thinking. If time's cycle stands for the immanence of law and time's arrow for the uniqueness of history, then Gould's dual career as a scientist and as a historian of science represents perhaps his greatest commitment to these two ways of understanding time."[31] Indeed, as Gerald Holton has so well explicated the principle, such themata are integral to the scientific process. Sulloway adds that such thematic pairs illuminate not only how science works but how the history of science operates, particularly in the works of Gould in his dual role as historian of science and scientific historian:

> Gould is one of those rare scientists who fully appreciates that the past is not always "just history" and that many problems in science cannot be conceptualized correctly unless one escapes the intellectual straitjacket of prevailing scientific mythologies. In this sense scientists are actually influenced by history all the time, even though they often disdain the subject as a waste of time. The textbook legends they fashion around their scientific heroes are value-laden visions of the world that often limit "the possibility of weighing reasonable alternatives," as Gould has emphasized about the history of geology. Thus doing the history of science is, for Gould at least, an essential part of doing good science.[32]

Doing good science is also an essential part of doing good history. The following excerpts from the essays provide an examplar for each of the five thematic dichotomies, demonstrating how Gould draws generalities out of minutiae.

Theory—Data. In an essay entitled "Bathybius and Eozoon," Gould explores the interaction between culture and science, and the relationship of concepts to percepts, in the context of a nineteenth-century debate over the nature of these two microscopic creatures that in time were revealed to be nothing more than geochemical by-products and thus were an embarrassing error to scientists. But historians know better, Gould explains: "They made sense in their own time; that they don't in ours is irrelevant. Our century is no standard for all ages; science is always an interaction of prevailing culture, individual eccentricity, and empirical constraint." The thematic lesson lies in the proper balance between theory and data, and how they play themselves out over time.

> Science contains few outright fools. Errors usually have their good reasons once we penetrate their context properly and avoid judgment according to our current perception of "truth." They are usually more enlightening than embarrassing, for they are signs of changing contexts. The best thinkers have the imagination to create organizing visions, and they are sufficiently adventurous (or egotistical) to float them in a complex world that can never answer "yes" in all detail. The study of inspired error should not engender a homily about the sin of pride; it should lead us to a recognition that the capacity for great insight and great error are opposite sides of the same coin—and that the currency of both is brilliance.[33]

Time's Arrow—Time's Cycle. In an essay entitled "Spin Doctoring Darwin," Gould pushes one of his favorite themes of change versus progress and bushes versus ladders in an analysis of how and why the Darwinian revolution has never been fully embraced, even today. Here he is not carping on creationists, but on evolutionary biologists who wrongly (in Gould's opinion) try to sneak back into Darwinism some higher purpose.

> Spin doctoring [Darwin] centers on two different subjects: the process of evolution as a theory and mechanism; and the pathway of evolution as a description of life's history. Spin doctoring for the process tries to depict evolution as inherently progressive, and as working toward some "higher" good in acting "for" the benefit of such groups as species or communities (not just for advantages of individual organisms), thereby producing such desired ends as harmonious ecosystems and well-designed organisms. Spin doctoring for the pathway reads the history of life as continuous flux with sensible directionality toward more complex and more brainy beings, thereby allowing us to view the late evolution of *Homo sapiens* as the highest stage, so far realized, of a predictable progress.[34]

Adaptationism—Nonadaptationism. In an essay entitled "Wallace's Fatal Flaw," Gould highlights the themes of optimality and suboptimality, continuity and discontinuity, by demonstrating how Alfred Russel Wallace erred in insisting (to Darwin's dismay) that natural selection could not account for the human mind because he could not conceive of an adaptive use for such a large organ during primate evolution. Therefore, Wallace reasoned, a higher intelligence must have intervened in the process, granting us such nonadaptive abilities as mathematics, music appreciation, and spiritual communication. This hyperadaptationism, in Gould's reading of the historical record, shows just how dangerous a scientific doctrine can become when carried to an extreme (and thus the lesson for today's hyperadaptationists).

> Natural selection may build an organ "for" a specific function or group of functions. But this "purpose" need not fully specify the capacity of that organ. Objects designed for definite purposes can, as a result of their structural complexity, perform many other tasks as well. A factory may install a computer only to issue the monthly pay checks, but such a machine can also analyze the election returns or whip anyone's ass (or at least perpetually tie them) in tic-tac-toe. Our large brains may have originated "for" some set of necessary skills in gathering food, socializing, or whatever; but these skills do not exhaust the limits of what such a complex machine can do.[35]

Punctuationism—Gradualism. In an essay entitled "The Interpretation of Diagrams," Gould considers the long-standing debate in geology over catastrophism versus uniformitarianism in the context of explaining the origins of the Cambrian "explosion" of life, arguing that the history of life from the beginning has been periodically punctuated by sudden and dramatic change (the "log phase" in the passage below) but most of the time remains relatively stable.

> The log phase of the Cambrian filled up the earth's oceans. Since then, evolution has produced endless variation on a limited set of basic designs. Marine life has been copious in its variety, ingenious in its adaptation, and (if I may be permitted an anthropocentric comment) wondrous in its beauty. Yet, in an important sense, evolution since the Cambrian has only recycled the basic products of its own explosive phase.[36]

Contingency—Necessity. In an essay entitled "The Horn of Triton," Gould uses the findings (and a striking photograph) from the *Voyager* spacecraft in its flyby of Neptune with its moon Triton, both in their crescent phases (thus the "horn") relative to the spacecraft, to consider what we can learn about the uniqueness of history versus the repeatability of nature's law's.

> I offer, as the most important lesson from *Voyager*, the principle of individuality for moons and planets. This contention should elicit no call for despair or surrender of science to the domain of narrative. We anticipated greater regularity, but have learned that the surfaces of planets and moons cannot be predicted from a few general rules. To understand planetary surfaces, we must learn the particular history of each body as an individual object—the story of collisions and catastrophes, more than steady accumulations; in other words, its unpredictable single jolts more than daily operations under nature's laws.[37]

These thematic pairs also help illuminate what is really going on in the so-called evolution wars.[38] When Gould, Lewontin, and Eldredge are pitted against Dawkins, Maynard Smith, and Dennett, it is almost always along a spectrum of one of these five themata. Maynard Smith's claim that Gould's ideas are confused and that he is giving nonprofessionals the wrong ideas about evolution is an indictment of Gouldian theory against others' data. Wright envisions a cyclical metaphor of time with directionality generating purpose, and thus is critical of Gould's emphasis on the directionless arrow in a purposeless cosmos. Ruse says that evolutionary biologists reject or ignore Gould's theory of punctuated equilibrium, but this is because he prefers phyletic gradualism. Dennett argues for a necessitating interpretation of the evolution of life whereas Gould emphasizes the contingent nature of history. Dawkins is a vocal defender of the adaptationist program in evolutionary theory, whereas Gould prefers to focus on the nonadaptive qualities of organisms. One wonders, in fact, if both sides in these various debates do not lean too close to the termini of each thematic pair as a corrective to the perceived exaggerated emphasis of the opponent on the other end of the spectrum. On the *Adaptationism—Nonadaptationism* theme, for example, Gould does not deny that natural selection creates well-adapted organisms. His point is that not everything in nature can be explained through the adaptationist paradigm.

> Darwinian theory is fundamentally about natural selection. I do not challenge this emphasis but believe that we have become overzealous about the power and range of selection by trying to

attribute every significant form and behavior to its direct action. In this Darwinian game, no prize is sweeter than a successful selectionist interpretation for phenomena that strike our intuition as senseless.[39]

Grandeur in This View of Life

This view of life is distinctly Gouldian in its struggle to find meaning in a contingently meaningless universe, to draw generalities out of the countless minutiae of the world, and to express it all in a literary style that balances professional scholarship with popular exposition. By deconstructing a single essay—"Modified Grandeur"—we see all of these elements neatly wrapped in one package, including biblical and literary references, history and philosophy of science and science studies, evolutionary theory, and several thematic pairs. Characteristically, Gould begins with an anecdote from his favorite pop-culture icon.

> In an old theatrical story, W. S. Gilbert was leading a rehearsal for the premiere of his most famous collaboration with A. S. Sullivan, *The Mikado*. At one point, Nanki-Poo learns that his beloved Yum-Yum is about to marry her guardian, Ko-Ko. Searching for a straw of light, he asks: "But you do not love him?" "Alas, no!" she replies. On hearing this sliver of mitigation, Nanki-Poo exclaims, "Rapture!"— or so Gilbert originally wrote. But at the rehearsal, Nanki-Poo stated his line too forcefully, given the limited comfort provided by Yum-Yum, and Gilbert shouted down from the balcony: "Modified rapture." The poor tenor, not grasping that Gilbert had only meant to correct his tone, and thinking instead that he had flubbed his line, exclaimed, "Modified rapture" at the reprise. This unintended correction elicited a good laugh, and so the line has remained ever since. If something so unvarnished as rapture must often be modified, let me pose a question in a similar vein: how shall we modify grandeur?[40]

The reference is to Darwin's final line from the *Origin of Species*: "There is grandeur in this view of life, with its several powers, having been originally breathed into a few forms or into one; and that, whilst this planet has gone cycling on according to the fixed law of gravity, from so simple a beginning endless forms most beautiful and most wonderful have been, and are being, evolved."[41] Darwin's denouement weaves together two themata—*Time's Arrow—Time's Cycle* and *Contingency—Necessity*—and Gould uses it to full rhetorical advantage when he then inquires why "evolved" appears only once in the *Origin* (the last word), and "evolution" never. Gould's answer buttresses his thematic preferences: "I believe that Darwin shunned this word because he recognized that natural selection, his theory of evolutionary mechanisms, contained no postulate about progress as a necessary feature of organic history—and, in vernacular English at the time, the word 'evolution' meant progress (literally, unfolding according to a preset plan)." Was this Darwin's intention, or Gould's interpretation? It is both. Gould uses the history of science to reinforce his thematic predilections, and vice versa, as is clear when he next chronicles the use of the word *grandeur* through three historical stages—(1) pre-Darwinian, (2) Darwinian, (3) post-Darwinian—while accenting their deeper meaning with a biblical reference: "We feel that we have gained greatly in factual and theoretical understanding through these three stages; but if we have lost a degree of grandeur for each step of knowledge gained, then we must fear Faust's bargain: 'For what shall it profit a man, if he shall gain the whole world, and lose his own soul?' (Mark 8:36)."[42]

1. Pre-Darwinian grandeur is that of the creator, an example of which can be found in Charles Bell's treatise "The Hand: Its Mechanism and Vital Endowments as Evincing Design" (which Gould read in preparation for the keynote address he delivered to the annual meeting of the American Society for Surgery of the Hand). "The Hand" became part of the famous *Bridgewater Treatises* "on the power, wisdom, and goodness of God, as manifested in the Creation." Bell wrote: "There is extreme grandeur in the thought of an anticipating or prospective intelligence: in reflecting that what was finally accomplished in man, was begun in times incalculably remote, and antecedent to the great revolutions which the earth's surface has undergone."[43] Gould notes in response: "What could be more grand, more extremely grand, than such a purposeful drama that puts *Homo sapiens* both in perpetual center stage and atop the ultimate peak at the end of the last act?"[44]

2. Darwinian grandeur contrasts time's cycle of planetary motion directed by necessitating laws with time's arrow of ever-changing, contingent history. "The 'grandeur in this view of life' lies squarely in the contrast

of cyclicity on a physical home with directionality of the biological inhabitants," Gould continues. "But Darwin has taken us down a peg, at least in terms of our standard cultural hopes and deep-seated arrogances. Bell's progressive creationism gave us a foreordained history of life, always perfect but moving upward toward an inevitable apotheosis in the origin of *Homo sapiens*. Darwin still sees expansion from original simplicity (the ever-ramifying tree of life in his metaphor), but specific outcomes are no longer ordained, and increase in complexity is only a broad trend, not a grand highway toward life's primary goal." Darwinian natural selection, says Gould as he slips in the *Adaptationism—Nonadaptationism* theme, "only produces adaptation to changing local environments, not global progress. A woolly mammoth is well-adapted to glacial climates, but cannot be called a generally improved elephant."[45]

However, says Gould, adding the *Theory—Data* theme into the mix, "Darwin was also a truly eminent Victorian—a wealthy, white male committed to (and greatly benefiting from) a society that had, perhaps more than any other in human history, made progress the centerpiece of its credo. How could Darwin jettison progress altogether in this age of industrial might, military triumph, and colonial expansion? Darwin therefore placed a modified form of progress back into his view of life through a supplementary argument about ecology and competition." This Darwin did by identifying two modes of the "struggle for existence": one against the physical environment and the other for limited resources. The first yields no progress, but the second can and does. Darwin writes: "The inhabitants of each successive period in the world's history have beaten their predecessors in the race for life, and are, in so far, higher in the scale of nature; and this may account for that vague yet ill-defined sentiment, felt by many paleontologists, that organization on the whole has progressed."[46]

3. Post-Darwinian grandeur, says Gould, falls on the nonprogressive and contingent end of the theme, with many paleontologists espousing the view that evolution contains no inherent progress within its processes and that if "the tape of life could be replayed from scratch" humans would be unlikely to arise again. "Bell exalted humans as the top rung of an inevitable ladder. Darwin perceived us as a branch on a tree, but still as a topmost shoot representing a predictable direction of growth. Many paleontologists, myself included, now view *Homo sapiens* as a tiny and unpredictable twig on a richly ramifying tree of life—a happy accident of the last

geological moment, unlikely ever to appear again if we could regrow the tree from seed."[47]

How are we to react to this loss of grandeur? Gould considers John Stuart Mill's suggestion that it is "better to be Socrates dissatisfied than a pig satisfied," but he rejects that and turns instead to the *Oxford English Dictionary*'s definition of grandeur: "transcendent greatness or nobility of intrinsic character." Finishing with a flourish Gould then ties together history, theory, and themata while admitting that ultimately there is an element of subjectivity in science: "For me then—and I will admit that grandeur must remain a largely personal and aesthetic concept—the modern view is grandest of all, for we have finally freed nature from primary judgment for placement of one little twig upon its copious bush. We can now step off and back—and see nature as something so vast, so strange (yet comprehensive), and so majestic in pursuing its own ways without human interference, that grandeur becomes the best word of all for expressing our interest, and our respect."[48]

We see in this 3,600-word essay the interaction of Gouldian history, theory, philosophy, and science, wrapped up in a tight literary package marketed to both professionals and the public. Gould is using the history of science to bolster his prejudices for certain theoretical interpretations of life's data—both biological and cultural—in support of the ends of the thematic pairs that best fit his worldview. As his critics are wont to point out, not all paleontologists accept this contingent and nonprogressive view of life, so Gould is building his case through every channel available. Gould is a historian and philosopher of science, but not intrinsically so. Yes, he is intensely passionate about "touching history" for its own sake, but this yearning is secondary to a larger purpose.[49] Not surprisingly, that purpose is very Darwinian.

Darwin's Dictum and Gould's Purpose

In 1861, less than two years after the publication of Charles Darwin's *Origin of Species*, in a session before the British Association for the Advancement of Science, a critic claimed that Darwin's book was too theoretical and that he should have just "put his facts before us and let them rest." In a letter to his friend Henry Fawcett, who was in attendance in his defense, Darwin explained the proper relationship between theory and data: "About thirty years ago there was much talk that geologists ought only to observe and not theorize; and I well remember someone saying that at this rate a man might as well go into a gravel-pit and count the pebbles and describe the colours. How odd it is that anyone should not see that all observation must be for or against some view if it is to be of any service!"[50]

The quotation is a favorite of Gould's, cited often in defense of his own philosophy of science that closely parallels that of Darwin. Gould's history of science, as well as his popular science expositions, is driven by this philosophy. In a two-part essay entitled "The Sharp-Eyed Lynx, Outfoxed by Nature,"[51] Gould shows how Galileo (the sixth member of the Academy of the Lynxes, a seventeenth-century organization dedicated to "reading this great, true, and universal book of the world," in the words of its founder Prince Federico Cesi) was outfoxed by the rings of Saturn for two reasons that tap directly into the *Theory—Data* theme: (1) Galileo's telescope was not powerful enough to clearly discern the structure of the rings; (2) Galileo had no model in his astronomy (or in his thoughts in general) for planetary rings. Given these conditions Galileo reported *Altissimum planetam tergeminum observavi*, "I have observed that the farthest planet is threefold" (in Gould's translation). Whenever the data of observation are unclear, the mind fills in the gaps. But if the mind has no model from which to work, imagination takes over, leading directly and powerfully to errors generated by expectation. Galileo could not "see" the rings of Saturn, either directly or theoretically, but he thought he could, and herein lies the problem, as Gould notes in Galileo's choice of words in his report: "He does not advocate his solution by stating 'I conjecture,' 'I hypothesize,' 'I infer,' or 'It seems to me that the best interpretation . . .' Instead, he boldly writes 'observavi'—I have observed. No other word could capture, with such terseness and accuracy, the major change in concept and procedure (not to mention ethical valuation) that marked the transition to what we call 'modern' science."[52] But this still is not Gould's deepest message in this essay, as it is still in the realm of a disconnected observation about the history of science. Gould brings it home to the reader:

> The idea that observation can be pure and unsullied (and therefore beyond dispute)—and that great scientists are, by implication,

people who can free their minds from the constraints of surrounding culture and reach conclusions strictly by untrammeled experiment and observation, joined with clear and universal logical reasoning—has often harmed science by turning the empiricist method into a shibboleth. The irony of this situation fills me with a mixture of pain for a derailed (if impossible) ideal and amusement for human foibles—as a method devised to undermine proof by authority becomes, in its turn, a species of dogma itself. Thus, if only to honor the truism that liberty requires eternal vigilance, we must also act as watchdogs to debunk the authoritarian form of the empiricist myth—and to reassert the quintessentially human theme that scientists can work only within their social and psychological contexts. Such an assertion does not debase the institution of science, but rather enriches our view of the greatest dialectic in human history: the transformation of society by scientific progress, which can only arise within a matrix set, constrained, and facilitated by society.[53]

Gould's purpose is Darwin's Dictum, presented in a popular genre for public and professional consumption and modified grandly to incorporate the greatest themata into this view of science as history and history as science.

Notes

5. Spin-Doctoring Science

1. Patrick Tierney, *Darkness in El Dorado: How Scientists and Journalists Devastated the Amazon* (New York: W. W. Norton, 2000).

2. Personal correspondence with Frank Miele, November 29, 2000.

3. Personal correspondence with Louise Brocket from W. W. Norton, November 20, 2000.

4. In defense of this statement see Michael Shermer, *How We Believe: The Search for God in an Age of Science* (New York: W. H. Freeman, 1999), chapter 7.

5. I coined "Darwin's Dictum" in my inaugural column for *Scientific American* (May 2001) from a letter Darwin wrote to a friend on September 18, 1861. The full quote reads: "About thirty years ago there was much talk that geologists ought only to observe and not theorize; and I well remember someone saying that at this rate a man might as well go into a gravel-pit and count the pebbles and describe the colours. How odd it is that anyone should not see that all observation must be for or against some view if it is to be of any service!"

6. Tierney, *Darkness in El Dorado*, p. 15. See also J. Lizot, *Tales of the Yanomami: Daily Life in the Venezuelan Forest* (New York: Cambridge University Press, 1985).

7. See http://www.anth.ucsb.edu/chagnon.html.

8. Sharon Begley, "Into the Heart of Darkness," *Newsweek* (November 27, 2000): 70–75.

9. Tierney, *Darkness in El Dorado*, p. 130.

10. Ibid., p. 131. Different authors use different spellings of the people's name, the two most common being Yanomamö and Yanomami. According to Chagnon, the "ö" is similar to the German "oe" and pronounced as it is in the German poet "Goethe."

11. Chagnon joked about the analogy with the film *The Gods Must Be Crazy* when he spoke at the Skeptics Society's 1996 Caltech conference.

12. Interview with Kenneth Good, December 5, 2000.

13. Kenneth Good, *Into the Heart: One Man's Pursuit of Love and Knowledge Among the Yanomami* (New York: HarperCollins, 1991), p. 5.

14. Interview with Napoleon Chagnon, December 12, 2000.

15. Derek Freeman, "Paradigms in Collision: Margaret Mead's Mistake and What It Has Done to Anthropology," *Skeptic* 5, no. 3 (1997): 66–73. This entire issue of *Skeptic* is devoted to anthropological controversies.

16. Interview with Bill Durham, December 1, 2000.

17. American Anthropological Association, *Code of Ethics*, from "Section A. Responsibility to people and animals with whom anthropological researchers work and whose lives and cultures they study."

18. Tierney, *Darkness in El Dorado*, pp. 132–33.

19. M. Roosevelt, "Yanomami: What Have We Done to Them?" *Time* (October 2, 2000): 77–78.

20. Interview with Napoleon Chagnon, December 12, 2000.

21. Statement by Kim Hill is at http://www.anth.ucsb.edu/chagnon.html.

22. Personal correspondence with Steven Pinker, December 1, 2000.

23. Go to http://www.anth.ucsb.edu to begin searching. Links and search engine scans under the names of the various anthropologists will net hundreds of pages of relevant documents.

24. Napoleon Chagnon, *Yanomamö* (New York: Harcourt Brace College Publishers, 1992), pp. xii–xiii.

25. Ibid., p. 7.

26. Ibid., p. 10.

27. Ibid., p. 206.

28. John Horgan, "The New Social Darwinists," *Scientific American* (October 1995): 150–57.

29. Napoleon Chagnon, "The Myth of the Noble Savage: Lessons from the Yanomamö People of the Amazon," presented at the Skeptics Society Conference on Evolutionary Psychology and Humanistic Ethics, March 30, 1996.

30. Ibid.

31. Ibid.

32. Ibid.

33. Interview with Donald Symons, November 28, 2000.

34. Interview with Napoleon Chagnon, December 12, 2000.

35. Good, *Into the Heart*, p. 42.

36. Interview with Kenneth Good, December 5, 2000.

37. Good, *Into the Heart*, p. 115.

38. Ibid., p. 116.

39. Ibid.

40. Chagnon, *Yanomamö*, p. 190.

41. Tierney, *Darkness in El Dorado*, p. 31.

42. Good, *Into the Heart*, p. 128.

43. Ibid., p. 129.

44. Ibid., p. 185.

45. Interview with Kenneth Good, December 5, 2000.

46. Ibid.

47. Interview with Napoleon Chagnon, December 12, 2000.

48. Chagnon, *Yanomamö*, p. 1.

49. Interview with Jared Diamond, November 27, 2000.

50. L. H. Keeley, *War Before Civilization: The Myth of the Peaceful Savage* (New York: Oxford University Press, 1996). A. Ferrill, *The Origins of War: From the Stone Age to Alexander the Great* (London: Thames and Hudson, 1988).

9. Exorcising Laplace's Demon

1. See chapter 10, "What If?" in this volume.

2. Michael Shermer, "The Chaos of History: On a Chaotic Model that Represents the Role of Contingency and Necessity in Historical Sequences," *Nonlinear Science 2*, no. 4 (1993): 1–13.

3. Personal correspondence, May 6, 1993.

4. Paul MacCready, "75 Reasons to Become a Scientist," *American Scientist* (September/October 1988): 457.

5. Personal correspondence, May 22, 1993. McNeill offered this additional caution: "I think the new burst of chaos theory has a lot to teach historians and am glad to find you doing it. In general we are an untheoretical profession: learn by apprenticeship and reflect little on the larger epistemological context of our inherited terms. But clarity is always desirable and you seem bent in that direction. I wish you well in illuminating the historical profession; but suspect most of my colleagues will not even try to understand!"

6. Niles Eldredge and Stephen Jay Gould. "Punctuated Equilibria: An Alternative to Phyletic Gradualism," in T. J. M. Schopf, ed., *Models in Paleobiology* (New York: Doubleday, 1972).

7. Lewis Binford, *An Archaeological Perspective* (New York: Academic Press, 1972); *In Pursuit of the Past: Decoding the Archaeological Record* (Berkeley, CA: University of California Press, 1983); *Working at Archaeology* (New York: Academic Press, 1983).

8. Fernand Braudel, *On History* (Chicago: University of Chicago Press, 1980).

9. C. G. Hempel, "The Function of General Laws in History," in P. Gardner, ed., *Theories of History* (New York: Free Press, 1959), p. 346.

10. Stephen Jay Gould, "Jove's Thunderbolts," *Natural History* (October 1994): 9.

11. Stephen Jay Gould, "The Horn of Triton," *Natural History* (December 1989): 18.

12. Ibid., p. 24.

13. Rom Harré, *The Principles of Scientific Thinking* (Chicago: University of Chicago Press, 1970).

14. Ian Stewart, *Does God Play Dice?* (London: Blackwell, 1990), p. 17.

15. Ilya Prigogine and Isabelle Stengers, *Order Out of Chaos* (New York: Bantam, 1984), p. 169.

16. Ibid., pp. 169–70.

17. Edward Lorenz, "Predictability: Does the Flap of a Butterfly's Wings in Brazil Set Off a Tornado in Texas?" Address at the AAAS annual meeting, Washington, D.C., December 29, 1979.

18. George Reisch, "Chaos, History, and Narrative," *History and Theory* 30, no. 1 (1991): 18.

19. For the historical sequence of QWERTY see Paul David's "Understanding the Economics of QWERTY: The Necessity of History," in W. N. Parker, ed., *Economic History and the Modern Economist* (New York: Blackwell, 1986). See also S. J. Gould's development of parallel biological and technological systems in "The Panda's Thumb of Technology," *Natural History* (January 1987): 14–23. For a general history of the typewriter, see F. T. Masi, ed., *The Typewriter Legend* (Secaucus, N.J.: Matsushita Electric Corporation

of America, 1985); F. J. Romano, *Machine Writing and Typesetting* (Salem: Gam Communications, 1986); and D. R. Hoke, *Ingenious Yankees: The Rise of the American System of Manufactures in the Private Sector* (New York: Columbia University Press, 1990). Hoke notes the paucity of historical records for reconstructing the history of the typewriter and was forced to rely on company histories, advertisements from magazines, photographs and illustrations of typewriters, surviving typewriters, and biographical material on the inventors, manufacturers, and entrepreneurs in the industry.

20. Per Bak and Kan Chen, "Self-Organized Criticality," *Scientific American* (January 1991): 46.

21. Ibid., p. 47

22. Ibid., p. 52

23. Ibid., p. 48

24. Stuart Kauffman, "Antichaos and Adaptation," *Scientific American* (February 1991): 78.

25. Ibid.

26. John Cohen and Ian Stewart, "Chaos, Contingency, and Convergence," *Nonlinear Science Today* 1, no. 2 (1991): 9–13.

27. Ibid., p. 3.

28. Ibid., pp. 2–3.

29. John L. Casti, *Complexification* (New York: HarperCollins, 1994), pp. 262–63.

30. Alan Macfarlane, *Witchcraft in Tudor and Stuart England* (London: HarperCollins, 1970).

31. Ibid.

32. Data from False Memory Syndrome Foundation, Philadelphia.

33. John Mack, *Abduction: Human Encounters with Aliens* (New York: Scribner, 1994). For a history of UFO sightings and alien abduction claims see Phil Klass, *UFO Abductions: A Dangerous Game* (Buffalo: Prometheus Books, 1988); and Robert Sheaffer, *The UFO Verdict: Examining the Evidence* (Buffalo: Prometheus Books, 1981).

34. The only scholarly paper ever done on the Mattoon story was by Donald Johnson, "The 'Phantom Anesthetist' of Mattoon," *Journal of Abnormal and Social Psychology* 40 (1945): 175–86. Willy Smith has written a skeptical analysis of Johnson's thesis and concludes that it was not a case of mass hysteria; rather, it was more likely a prankster and/or journalistic scam. "The Mattoon Phantom Gasser: The Definitive Analysis 50 Years Later," *Skeptic* 3, no. 1 (1994). Other works that document the rise and fall of social movements in the pattern modeled in this paper include: Philip Jenkins, *Intimate Enemies: Moral Panics in Contemporary Great Britain* (Hawthorne, N.Y.: Aldine de Gruyter, 1992); and James Richardson, Joel Best, and David Bromley, eds., *The Satanism Scare* (Hawthorne, N.Y.: Aldine de Gruyter, 1991).

35. Quoted in Philipp Frank, *Philosophy of Science: The Link Between Science and Philosophy* (Englewood Cliffs, N.J.: Prentice-Hall, 1957).

36. Herman Melville, *Moby-Dick* (Indianapolis: Bobbs-Merrill,1964), pp. 135–36.

10. What If?

1. Michael Shermer, "The Chaos of History," *Nonlinear Science Today* 2, no. 4 (1993): 1–13.

2. Michael Shermer, "The Crooked Timber of History," *Complexity* 2, no. 6 (1997): 23–29.

3. Michael Shermer, *Denying History* (Berkeley: University of California Press, 2000).

4. R. J. Van Pelt and D. Dwork, *Auschwitz* (New York: W. W. Norton, 1996).

5. R. J. Van Pelt, "A Site in Search of a Mission," in Y. Gutman and M. Berenbaum, eds., *Anatomy of the Auschwitz Death Camp* (Bloomington: Indiana University Press, 1994).

6. Ibid., p. 94.

7. Van Pelt and Dwork, *Auschwitz*, p. 11.

8. Ibid., pp. 150–51.

9. Ibid., p. 11.

10. J. Bulhof, "What If? Modality and History," *History & Theory* 38, no. 2 (1999): 145–68.

11. D. Goldhagen, *Hitler's Willing Executioners* (New York: Knopf, 1996).

12. Michael Shermer, "We Are the World: A Review of *Nonzero* by Robert Wright," *Los Angeles Times Book Review*, February 6, 2000.

13. Robert Wright, *Nonzero: The Logic of Human Destiny* (New York: Pantheon, 2000).

14. Ibid., p. 317.

15. Ibid., p. 7.

16. Ibid., p. 291.

17. Ibid., p. 292.

18. Ibid., pp. 292–93.

19. R. G. Klein, *The Human Career* (Chicago: University of Chicago Press, 1999), pp. 367–493.

20. Ibid., p. 477.

21. Richard Leakey, *The Origin of Humankind* (New York: Basic Books, 1994), p. 134.

22. Klein, *Human Career*, pp. 441–42.

23. C. Wills, *Children of Prometheus* (Reading, Mass.: Perseus Books, 1998), pp. 143–45.

24. Michael Shermer, *How We Believe* (New York: W. H. Freeman, 1999).

25. Klein, *Human Career*, p. 469.

26. Ian Tattersall, "Once We Were Not Alone," *Scientific American* (January 2000): 56–62.

27. Ian Tattersall, *The Fossil Trail* (New York: Oxford University Press, 1995), p. 212.

28. Leakey, *Origin of Humankind*, p. 132.

29. Ibid., p. 138.

30. Ibid., p. 20.

31. Tattersall, *Fossil Trail*, p. 246.

32. Ibid., p. 62.

33. N. Roberts, *The Holocene* (Oxford, UK: Basil Blackwell, 1989).

34. Jared Diamond, *Guns, Germs, and Steel* (New York: W. W. Norton, 1997).

35. Ibid., pp. 424–25.

36. Michael Shermer, "Humans, History, and Environments: An Interview with Jared Diamond," *Skeptic* 8, no. 3 (2000): 41–47.

14. This View of Science

1. John Brockman, *The Third Culture: Beyond the Scientific Revolution* (New York: Simon and Schuster, 1995).

2. William Poundstone, *Carl Sagan: A Life in the Cosmos* (New York: Henry Holt, 1999), pp. 261–62. Keay Davidson, *Carl Sagan: A Life* (New York: Wiley, 1999), pp. 331–33.

3. Cited on the National Science Foundation Web site: www.nsf.gov/sbe/srs/seind00access/c8/c8s4.htm. The relevant text reads: "One of the most frequently cited reasons for scientists' reluctance to talk to the press is the so-called Carl Sagan effect, that is, renowned scientist Carl Sagan was criticized by his fellow scientists who assumed that because Sagan was spending so much time communicating with the public, he must not have been devoting enough time to his research."

4. Poundstone, Carl Sagan, pp. 112, 357; Davidson, Carl Sagan, pp. 202–205, 389–392. Poundstone describes the debate at the NAS over Sagan's nomination this way: "Texas A&M chemist Albert Cotton took dead aim on the popularization issue. He judged popularization to be oversimplification—symptomatic of an inadequacy in doing science. There were nods of approval. Rosalyn Yalow, the Nobel-laureate medical physicist, shook her head, vowing, 'Never, never.' One foe said that the fact that Carl Sagan had even gotten on the ballot demonstrated how 'dangerous' it was to allow open nominations" (p. 357).

5. Michael White, "Eureka! They Like Science," The Sunday Times, December 13, 1992.

6. Awards include a National Book Award for The Panda's Thumb, a National Book Critics Circle Award for The Mismeasure of Man, the Phi Beta Kappa Book Award for Hen's Teeth and Horse's Toes, and a Pulitzer Prize Finalist for Wonderful Life, on which Gould commented, "Close but, as they say, no cigar." Forty-four honorary degrees and sixty-six major fellowships, medals, and awards bear witness to the depth and scope of his accomplishments in both the sciences and humanities. He even has a Jupiter-crossing asteroid named after him ("Stephengould," as by IAU convention), discovered by Gene Shoemaker in 1992. Awards and citations taken from Gould's curriculum vitae, dated September 2000. The reference to Gould as "America's evolutionist laureate" appears in numerous publications, but first appears, ironically, in Robert Wright's highly critical review of Wonderful Life, in The New Republic (January 29, 1990). He meant it sarcastically, but it has been adopted since in praise.

7. Bernard D. Davis, Storm over Biology: Essays on Science, Sentiment, and Public Policy (Buffalo: Prometheus Books, 1986), pp. 130, 136.

8. Daniel Dennett, Darwin's Dangerous Idea (New York: Simon and Schuster, 1995), pp. 262–312.

9. Richard Dawkins, Unweaving the Rainbow: Science, Delusion and the Appetite for Wonder (Boston: Houghton Mifflin, 1998), pp. 197–98.

10. John Maynard Smith, "Genes, Memes, and Minds," New York Review of Books 42, no. 19 (1995): 46. The quotation is repeated often by Gould's critics: Richard Dawkins (n. 9), John Alcock (n. 11), Robert Wright (n. 12), and Michael Ruse (n. 13). Gould replied in the New York Review of Books 44, no. 10 (1997): 34–37:

> He really ought to be asking himself why he has been bothering about my work so intensely, and for so many years. Why this dramatic change? Has he been caught up in apocalyptic ultra-Darwinian fervour? I am, in any case, saddened that his once genuinely impressive critical abilities seem to have become submerged within the simplistic dogmatism epitomized by Darwin's Dangerous Idea, a dogmatism that threatens to compromise the true complexity, subtlety (and beauty) of evolutionary theory and the explanation of life's history.

11. John Alcock, "Unpunctuated Equilibrium in the Natural History Essays of Stephen Jay Gould," Evolution and Human Behavior 19 (1998): 321–35.

12. Robert Wright, "The Intelligence Test: A Review of Wonderful Life: The Burgess Shale and the Nature of History by Stephen Jay Gould," The New Republic (January 29,

1990): 32. See also Robert Wright, "Homo deceptus," *Slate* (www.slate.com) (November 27, 1996); Robert Wright, "The Accidental Creationist," *The New Yorker* (December 13, 1999): 56; and Robert Wright, *Nonzero: The Logic of Human Destiny* (New York: Pantheon, 2000).

13. Michael Ruse, *The Evolution Wars: A Guide to the Debates* (Denver: ABC-CLIO, 2000), pp. 247–48.

14. Ronald Numbers, in response to a questionnaire about Gould's strengths and weaknesses as a scientist, conducted in June 2000, as part of a larger survey to assess the personality characteristics of eminent scientists.

15. Charles Darwin to Henry Fawcett, September 18, 1861, letter number 133 in F. Darwin, ed., *More Letters of Charles Darwin*, vol. 1 (New York: D. Appleton, 1903), pp. 194–96.

16. This rough classification of Gould's books includes: natural history (*Illuminations, Crossing Over*, and the nine essay collections); history of science/science studies (*The Mismeasure of Man, An Urchin in the Storm, Finders Keepers, Questioning the Millennium, Rocks of Ages*); evolutionary theory (*Ontogeny and Phylogeny, The Book of Life, Full House, The Structure of Evolutionary Theory*); paleontology/geology (*Time's Arrow, Time's Cycle; Wonderful Life*). In developing the more elaborate taxonomic system for classifying Gould's scientific and scholarly writings a number of specialists were consulted, including paleontologist Donald Prothero from Occidental College, historian of science Frank Sulloway from UC Berkeley, science historian and *Natural History* magazine editor Richard Milner, and, of course, Gould himself, who was patient through my numerous taxonomic queries.

17. Stephen Jay Gould, "Entropic Homogeneity Isn't Why No One Hits .400 Anymore," *Discover* (August 1986): 60–66. Gould, "Phyletic Size Decrease in Hershey Bars" in Charles J. Rubin et al., eds., *Junk Food* (New York: The Dial Press/James Wade, 1980), pp. 178–79. Gould, "Mickey Mouse Meets Konrad Lorenz," *Natural History* (May 1997): 30–36.

18. See, for example, the prologue to Stephen Jay Gould, *Ever Since Darwin* (New York: W. W. Norton, 1977), describing the essays: "They range broadly from planetary and geological to social and political history, but they are united (in my mind at least) by the common thread of evolutionary theory—Darwin's version. I am a tradesman, not a polymath; what I know of planets and politics lies at their intersection with biological evolution" (pp. 13–14). He repeats the line in the prefaces to *The Panda's Thumb* (New York: W. W. Norton, 1980, p. 12) and *Dinosaur in a Haystack* (New York: W. W. Norton, 1995), p. ix.

19. Gould, *Dinosaur in a Haystack*, pp. xiii–xiv.

20. Stephen Jay Gould, "The Streak of Streaks," *The New York Review of Books* 35 (1988): 8–12. The comparison of Gould's essay streak to DiMaggio's hitting streak was made by former major league baseball player Bruce Bochte in an introduction of Gould for a talk at the Academy of Natural Sciences in San Francisco, recounted by Gould in *Eight Little Piggies* (New York: W. W. Norton, 1993), p. 11.

21. Gould's first essay in *Natural History* was preceded in March 1973 with a standalone article on "The Misnamed, Mistreated, and Misunderstood Irish Elk." He explains the origin of the column in the preface to *The Lying Stones of Marrakech* (New York: Harmony, 2000): "In the fall of 1973, I received a call from Alan Ternes, editor of *Natural History* magazine. He asked me if I would like to write columns on a monthly basis, and he told me that folks actually get paid for such activities. (Until that day, I had published only in technical journals.) The idea intrigued me, and I said that I'd try three or four. Now, 290

monthly essays later (with never a deadline missed), I look only a little way forward to the last item of this extended series—to be written, as number 300 exactly, for the millennial issue of January 2001" (p. 1).

22. Gould even commented on this trend in a parenthetical note in the prologue to *The Flamingo's Smile* (New York: W. W. Norton, 1985): ". . . my volumes have become progressively longer for an unchanging number of essays—a trend more regular than my mapped decline of batting averages from essay 14, and a warning signal of impending trouble if continued past a limit reached, I think, by this collection" (p. 15). Trouble or not, the length stretched by another thousand words per essay by the mid-1990s.

23. Alcock, "Unpunctuated Equilibrium," pp. 322–23.

24. The Marx reference comes from Gould, "The Horn of Triton," *Natural History* (December 1989): 18. Not included in the count of biblical references were passages from essays devoted entirely to biblical exegesis, which include: "Fall in the House of Usher" (November 1991); "The Pre-Adamite in a Nutshell" (November 1999); "The First Day of the Rest of Our Life (April 2000); and "The Narthex of San Marco and the Pangenetic Paradigm" (July/August 2000).

25. Stephen Jay Gould, "In Touch with Walcott," *Natural History* (July 1990): 6–12.

26. Richard Milner, "Stephen Jay Gould Is My Name," based on "My Name Is John Wellington Wells" from *The Sorcerer*, performed on October 7, 2000, at the Festschrift held in Gould's honor at the California Institute of Technology, copyright © Richard Milner, 2001, reprinted by permission.

27. I began reading Gould's essays in 1985 starting with the essay collections. After that I read most of the essays in their original publication in *Natural History*, and reread many of them when they were republished in book form. Finally, in late 2000 I went through all three hundred essays in chronological order, page by page, in order to classify them in this taxonomic scheme. It soon became clear that for most of the essays there were multiple layers of literary, scientific, and philosophical complexity, so I developed this three-tiered system to discern the larger patterns. When it became apparent that in most of the essays there was also a strong historical element, I added another three-tiered division to classify the relevant essays by their historical subject or theme. My coding scheme was developed on a handful of randomly selected essays to the point where it became relatively obvious what the primary, secondary, and tertiary themes were in each. I then went through the entire corpus sequentially. There is a certain amount of subjectivity to the process, but knowing Gould's essays as well as I do, I can say with confidence that there would be little dispute of my coding outcomes. Readers can obtain a copy of the raw data by e-mail at skepticmag@aol.com.

28. Frank Sulloway was invaluable in helping classify Gould's essays in this complex network of literary taxonomy, particularly with regard to the relationship of the history and philosophy of science in Gould's work.

29. Stephen Jay Gould, "Leonardo's Living Earth," *Natural History* (May 1997): 16–22. For many of the seventy-six historical biographies, of course, Gould relied on secondary sources for general information about the individual, but for almost all of them he turned to primary documents, especially those composed by the subjects themselves. In many instances this meant reading historical Latin, French, German, Russian, and other languages that Gould had to teach himself in order to avoid the risk of relying on others' translations.

30. The total comes to 379 because a number of essays had two dominant themata, and three had three deep themes: "Modified Grandeur," *Natural History* (March 1993); "Spin Doctoring Darwin," *Natural History* (July 1995); and "What Does the Dreaded 'E' Word Mean, Anyway?" *Natural History* (February 2000).

31. Frank Sulloway, "The Metaphor and the Rock: A Review of *Time's Arrow, Time's Cycle: Myth and Metaphor in the Discovery of Geological Time* by Stephen Jay Gould," *New York Review of Books* (May 2, 1987): 37–40. Sulloway notes Gerald Holton's important contributions to understanding the role of such themata in the development of all scientific ideas: "Gerald Holton has argued that all science is inspired by such bipolar 'themata,' which transcend the strictly empirical character of science by giving a primary role to human imagination." See Gerald Holton, *The Thematic Origins of Scientific Thought: Kepler to Einstein* (Cambridge: Harvard University Press, 1988).

32. Sulloway, "The Metaphor and the Rock," p. 39.

33. Stephen Jay Gould, "Bathybius and Eozoon," *Natural History* (April 1978): 16–22.

34. Stephen Jay Gould, "Spin Doctoring Darwin," *Natural History* (July 1995): 12–18.

35. Stephen Jay Gould, "Wallace's Fatal Flaw," *Natural History* (January 1980): 26–40.

36. Stephen Jay Gould, "The Interpretation of Diagrams," *Natural History* (August/September 1976): 18–28.

37. Stephen Jay Gould, "The Horn of Triton," *Natural History* (December 1989): 18–27.

38. See Ruse, *The Evolution Wars*; Ullica Segerstrale, *Defenders of the Truth: The Battle for Science in the Sociobiology Debate and Beyond* (Oxford: Oxford University Press, 2000); Richard Morris, *The Evolutionists: The Struggle for Darwin's Soul* (New York: W. H. Freeman, 2001).

39. Stephen Jay Gould, "Only His Wings Remained," *Natural History* (September 1984): 10–18.

40. Stephen Jay Gould, "Modified Grandeur," *Natural History* (March 1993): 14–20.

41. This wording comes from the first edition of the *Origin*. In later editions Darwin added this modifying clause (noted in italics in original): "*. . . having been originally breathed by the Creator into a few forms or into one.*"

42. Gould, "Modified Grandeur," p. 14.

43. Quoted in ibid., p. 15.

44. Ibid., p. 18.

45. Ibid.

46. Quoted in ibid.

47. Ibid., p. 20.

48. Ibid.

49. Gould added this reflective comment on the essay: "I was also making a little joke about adjectives, from Bell's 'extreme' grandeur to Darwin's plain and unmodified 'grandeur' to the truly 'modified' grandeur—hence the Gilbert and Sullivan intro story—of contingency which, ironically in a non–spin doctored conceptual sense is, after all, the greatest grandeur of all because it is most contrary to our hopes and expectations and therefore forces us to think!" (personal communication, May 15, 2001).

50. Darwin to Henry Fawcett, September 18, 1861, *More Letters of Charles Darwin*, vol. I. Of the final clause of the line, "if it is to be of any service," Gould commented light-heartedly: "It tickles me that the quote has six words in a row with only two letters each. Now this must be rare! (but how to measure it??)" (personal communication, May 15, 2001).

51. Stephen Jay Gould, "The Sharp-Eyed Lynx, Outfoxed by Nature," *Natural History* (May 1998): 16–21, 70–72.

52. Ibid., p. 18.

53. Ibid., p. 19.

Permissions

Essay Credits

1. "Psychic for a Day: Or, How I Learned Tarot Cards, Palm Reading, Astrology, and Mediumship in Twenty-four Hours." Originally published in *Skeptic* 10, no. 1 (2003).

4. "The Virtues of Skepticism: A Way of Thinking and a Way of Life." Originally published as "Let Us Reflect: How a Thoughtful, Inquiring Watchman Provided a Mark to Aim at" in Paul Kurtz, ed. *25 Years of Skepticism* (Amherst, Mass.: Prometheus Books, 2001).

5. "Spin-Doctoring Science: Science as a Candle in the Darkness of the Anthropology Wars." Originally published in *Skeptic* 9, no. 1 (2000).

6. "Psyched Up, Psyched Out: Can Science Determine if Sports Psychology Works?" Originally published in *Scientific American Presents: Building the Elite Athlete—The Science and Technology of Sport* (fall 2000).

7. "Shadowlands: Science and Spirit in Life and Death." Originally published in *Science and Spirit* (summer 2004).

9. "Exorcising Laplace's Demon: Clio, Chaos, and Complexity." Originally published in *History and Theory* 34, no. 1 (1995).

10. "What If?: Contingencies and Counterfactuals: What Might Have Been and What Had to Be." Originally published in *Skeptic* 8, no. 3 (2000).

12. "History's Heretics: Who and What Mattered in the Past?" Originally published in *Skeptic* 2, no. 1 (1993). Updated in 2000.

13. "The Hero on the Edge of Forever: Gene Roddenberry, *Star Trek*, and the Heroic in History." Originally published in *Skeptic* 3, no. 1 (1994).

14. "This View of Science: The History, Science, and Philosophy of Stephen Jay Gould." Originally published in *Social Studies of Science* (September 2002) and revised for this collection.

Illustration Credits

Figure I.1. Illustration by Pat Linse.

Figure I.2. Rendered by Pat Linse.

Figure I.3. Rendered by Pat Linse.

Figures I.4, 1.5, I.6. Courtesy of Jerry Andrus.

Figures I.7, 1.8, I.9. Figure I.7 taken from *Viking Orbiter 1*, 1976. Figure I.8 taken from *Mars Global Surveyor*, 2001. Figures I.9a and I.9b taken in 1999. Courtesy of NASA.

Figures I.10, I.11. Author's collection.

Figure I.12. Illustrations by Pat Linse.

Figure 1.1. Photograph by Pat Linse.

Figure 1.2. Graphic by Pat Linse.

Figure 1.3. Rendered by Pat Linse from an Internet site.

Figure 1.4. Rendered by Pat Linse from an Internet site.

Figure 5.1. From Kenneth Good, *Into the Heart* (New York: HarperCollins,1991), facing p. 140. Courtesy of Kenneth Good.

Figure 5.2. Courtesy of Napoleon Chagnon.

Figure 6.1. Author's collection.

Figure 7.1. Author's collection.

Figure 8.1. From a private collection.

Figure 8.2. From the Museum of Art, the University of Melbourne.

Figure 8.3. From the Museum of Art, the University of Melbourne.

Figure 8.4. National Library of Australia.

Figure 9.1. Author's collection.

Figures 9.2, 9.3, 9.4, 9.5. Graphics by Pat Linse.

Figure 10.1. Illustration by Pat Linse.

Figure 10.2. Redrawn by Pat Linse, from R. J. Van Pelt and D. Dwork, *Auschwitz* (New York: W. W. Norton, 1996).

Figure 10.3 From Ian Tattersall, *The Fossil Trail* (New York: Oxford University Press, 1995).

Figure 10.4. From Jared Diamond, *Guns, Germs, and Steel* (New York: W. W. Norton, 1997).

Figures 11.1, 11.2, 11.3. Illustrations by Pat Linse.

Figure 12.1. Detail from the engraving *Histoire de l'origine et des premiers progres de l'imprimerie* by Prosper Marchand. From Elizabeth L. Eisenstein, *The Printing Revolution in Early Modern Europe* (New York: Cambridge University Press, 1993), frontispiece.

Figure 12.2. Illustration by Johannes Stradanus (1523–1605). From the British Museum.

Figure 12.3. From the British Museum.

Figure 12.4. Frontispiece from Francis Bacon's 1620 *Instauratio Magna*. From Elizabeth L. Eisenstein, *The Printing Revolution in Early Modern Europe* (New York: Cambridge University Press, 1993), p. 258.

Figure 14.1. Painting by Pat Linse.

Figures 14.2, 14.3, 14.4, 14.5, 14.6, 14.7, 14.8. Graphs by Pat Linse.

Acknowledgments

As I have in all of my books, I would like to acknowledge a number of individuals who have contributed not only to this book but to my work in general, starting with my agents Katinka Matson and John Brockman, not only for their personal support of my work but for what they have done to help shape the genre of science writing into a "third culture" on par with other cultural traditions. And thanks as well to Paul Golob at Henry Holt/Times Books, who oversaw the project, and especially to Robin Dennis, my editor, who has greatly shaped my thinking and writing into a finer prose than I otherwise would have produced. For this book, Robin also helped choose the most appropriate essays and articles to contribute, a selection process that only an unbiased mind could have implemented. I also acknowledge Muriel Jorgensen for constructive copyediting of the manuscript, Lisa Fyfe for the creative cover design, Victoria Hartman for interior design, and Chris O'Connell for overall production. Kate Pruss in the Holt publicity department has unfailingly supported our long-range mission of promoting science and critical thinking by reaching larger audiences, and for this I am deeply grateful.

Since many of the chapters in this book were originally published in

Skeptic magazine, special thanks go to art director Pat Linse for her important contributions in preparing the illustrations, graphs, and charts for this and my other works, as well as for her unmitigated and deeply appreciated friendship and support. The skeptical movement in general owes a debt of gratitude to Pat for her many behind-the-scenes actions that have irrevocably shaped modern skepticism into a viable social movement. She is first among equals in the pantheon of those most influential in skepticism, and this is reflected in the dedication of this book.

I also wish to recognize the office staff of the Skeptics Society and *Skeptic* magazine, including Matthew Cooper, Tanja Sterrmann, and Daniel Loxton; as well as senior editor Frank Miele; senior scientists David Naiditch, Bernard Leikind, Liam McDaid, and Thomas McDonough; contributing editors Tim Callahan, Randy Cassingham, Clayton Drees, Steve Harris, Tom McIver, Brian Siano, and Harry Ziel; editorial assistants Gene Friedman, Sara Meric, and the late Betty McCollister; and photographer David Patton and videographer Brad Davies for their visual record of the Skeptics' Caltech Science Lecture Series. I would also like to recognize *Skeptic* magazine's board members: Richard Abanes, David Alexander, the late Steve Allen, Arthur Benjamin, Roger Bingham, Napoleon Chagnon, K. C. Cole, Jared Diamond, Clayton J. Drees, Mark Edward, George Fischbeck, Greg Forbes, the late Stephen Jay Gould, John Gribbin, Steve Harris, William Jarvis, Lawrence Krauss, Gerald Larue, William McComas, John Mosley, Richard Olson, Donald Prothero, James Randi, Vincent Sarich, Eugenie Scott, Nancy Segal, Elie Shneour, Jay Stuart Snelson, Julia Sweeney, Frank Sulloway, Carol Tavris, and Stuart Vyse.

Thanks for institutional support for the Skeptics Society at the California Institute of Technology go to David Baltimore, Susan Davis, Chris Harcourt, and Kip Thorne. Larry Mantle, Ilsa Setziol, Jackie Oclaray, Julia Posie, and Linda Othenin-Girard at KPCC 89.3 FM radio in Pasadena have been good friends and valuable supporters of promoting science and critical thinking on the air. Thanks to Linda Urban at Vroman's bookstore in Pasadena for her support; Robert Zeps and John Moores have been especially supportive of both the Skeptics Society as well as the skeptical movement in America. Finally, special thanks go to those who help at every level of our organization: Stephen Asma, Jaime Botero, Jason Bowes, Jean Paul Buquet, Adam Caldwell, Bonnie Callahan, Tim Callahan, Cliff

Caplan, Randy Cassingham, Shoshana Cohen, John Coulter, Brad Davies, Janet Dreyer, Bob Friedhoffer, Michael Gilmore, Tyson Gilmore, Andrew Harter, Diane Knudtson, Joe Lee.

I would also like to acknowledge John Rennie and Mariette DiChristina at *Scientific American* for providing skepticism a monthly voice that reaches so many people. I look forward each month to writing my column more than just about anything else I do in my working life.

Finally, I thank my daughter, Devin, for bringing so much joy just by being herself and my wife, Kim, with whom I have now shared life for nearly two decades.

Index

Entries in *italics* refer to figures, tables and illustrations.

About the Author

DR. MICHAEL SHERMER is the founding publisher of *Skeptic* magazine (www.skeptic.com), the director of the Skeptics Society, a contributing editor of and monthly columnist for *Scientific American*, the host of the Skeptics Lecture Series at Caltech, and the cohost and producer of the thirteen-hour Fox Family television series *Exploring the Unknown*. He is the author of a trilogy of books on human belief, *Why People Believe Weird Things*, *How We Believe*, and *The Science of Good and Evil*. His other works include *In Darwin's Shadow*, about the life and science of the codiscoverer of natural selection Alfred Russel Wallace; *The Borderlands of Science*, about the fuzzy land between science and pseudoscience; and *Denying History*, on Holocaust denial and other forms of historical distortion. He is also the general editor of *The Skeptic Encyclopedia of Pseudoscience*.

Dr. Shermer received his B.A. in psychology from Pepperdine University, his M.A. in experimental psychology from California State University, Fullerton, and his Ph.D. in the history of science from Claremont Graduate University. He lives in southern California.